Lecture Notes in Computer Science 15399

Founding Editors

Gerhard Goos

Juris Hartmanis

W0234430

The series Lecture Notes in Computer Science (LNCS), including its subseries Lecture Notes in Artificial Intelligence (LNAI) and Lecture Notes in Bioinformatics (LNBI), has established itself as a medium for the publication of new developments in computer science and information technology research, teaching, and education.

LNCS enjoys close cooperation with the computer science R & D community, the series counts many renowned academics among its volume editors and paper authors, and collaborates with prestigious societies. Its mission is to serve this international community by providing an invaluable service, mainly focused on the publication of conference and workshop proceedings and postproceedings. LNCS commenced publication in 1973.

Fangguo Zhang · Weiwei Lin · Hongyang Yan
Editors

Artificial Intelligence Security and Privacy

Second International Conference, AIS&P 2024
Guangzhou, China, December 6–7, 2024
Proceedings

 Springer

Editors
Fangguo Zhang
Sun Yat-sen University
Guangzhou, China

Weiwei Lin
South China University of Technology
Guangzhou, China

Hongyang Yan
Guangzhou University
Guangzhou, China

ISSN 0302-9743   ISSN 1611-3349 (electronic)
Lecture Notes in Computer Science
ISBN 978-981-96-1147-8   ISBN 978-981-96-1148-5 (eBook)
https://doi.org/10.1007/978-981-96-1148-5

Preface

The Second International Conference on Artificial Intelligence Security and Privacy (AIS&P 2024) was held in Guangzhou, China during December 6–7, 2024. AIS&P serves as an international conference for researchers to exchange the latest research progress in all areas such as artificial intelligence, security and privacy, and their applications. This volume contains papers presented at AIS&P 2024.

The conference received 47 submissions. The committee accepted 14 regular papers to be included in the conference program. Every paper received 2 or 3 reviews. The proceedings contain revised versions of the accepted papers. While revisions were expected to take the referees' comments into account, this was not enforced and the authors bear full responsibility for the content of their papers.

AIS&P 2024 was organized by Guangzhou University. The conference would not have been such a success without the support of the University, and we sincerely thank the staff for their continued assistance and support.

We would also like to thank the authors who submitted their papers to AIS&P 2024, and the conference attendees for their interest and support. We thank the Organizing Committee for their time and effort dedicated to arranging the conference. This allowed us to focus on the paper selection and deal with the scientific program. We thank the Program Committee members and the external reviewers for their hard work in reviewing the submissions; the conference would not have been possible without their expert reviews. Finally, we thank the EasyChair system and its operators, for making the entire process of managing the conference convenient.

October 2024

Fangguo Zhang
Weiwei Lin
Hongyang Yan

Organization

General Chairs

Fangguo Zhang	Sun Yat-sen University, China
Weiwei Lin	South China University of Technology, China
Hongyang Yan	Guangzhou University, China

Program Chairs

Jaideep Vaidya	Rutgers University, USA
Moncef Gabbouj	Tampere University, Finland
Jin Li	Guangzhou University, China
Cong Wang	City University of Hong Kong, China
Lei Chen	Hong Kong University of Science and Technology, China
Hongmin Liu	University of Science and Technology Beijing, China

Track Chairs

Muhammad Khurram Khan	King Saud University, Saudi Arabia
Yun Peng	Guangzhou University, China
Kwangjo Kim	KAIST, South Korea
Shaowei Wang	Guangzhou University, China

Publication Chairs

Weizhi Meng	Technical University of Denmark, Denmark
Francesco Palmieri	University of Salerno, Italy

Publicity Chairs

Yu Wang	Guangzhou University, China
Anli Yan	Guangzhou University, China

Steering Committee

Albert Zomaya	University of Sydney, Australia
Jaideep Vaidya	Rutgers University, USA
Moncef Gabbouj	Tampere University, Finland
Jin Li	Guangzhou University, China

Contents

BadHAR: Backdoor Attacks in Federated Human Activity Recognition Systems

Dongping Zhang[1], Bing Mi[2(✉)], and Kongyang Chen[1,3(✉)]

[1] School of Artificial Intelligence, Guangzhou University, Guangzhou, China
[2] Guangdong University of Finance and Economics, Guangzhou, China
mibing89@gmail.com
[3] Pazhou Lab, Guangzhou, China
kychen@gzhu.edu.cn

Abstract. With the rapid development of Internet of Things (IoT) technology, Human Activity Recognition (HAR) has seen increasingly widespread applications in daily life, such as in health monitoring and activity tracking features commonly found in smartphones and smartwatches. However, despite the significant advancements in preserving user privacy through federated learning, numerous challenges remain, particularly in terms of security. During the training process in federated learning, the data on participating client devices is not visible to the server, making the system vulnerable to attacks by malicious clients. Specifically, a malicious client may use backdoor-infused data to train its local model, which can then compromise the global model when the server aggregates these models. To the best of our knowledge, no existing research has addressed the issue of backdoor attacks in Federated Human Activity Recognition. To address this gap, we propose an effective method for generating HAR backdoor data based on observations from public datasets. Experimental results indicate that our backdoor attack method achieves excellent attack success rates across six public HAR datasets. Our work offers new insights and potential solutions for enhancing the security of HAR models in federated learning.

Keywords: Federated Learning · Human Activity Recognition · Backdoor Attack

1 Introduction

The objective of Human Activity Recognition (HAR) is to identify and understand human activities by analyzing information gathered from various data sources [1,2]. HAR is generally categorized into two types: vision-based HAR and sensor-based HAR. The former relies on cameras and image processing techniques, while the latter utilizes sensor data from devices, such as accelerometers, gyroscopes, and magnetometers. With the rapid advancement of semiconductor technology, modern sensor technology has significantly improved in both accuracy and reliability, enabling the provision of higher-quality data. Simultaneously, the proliferation of low-power sensors and wearable devices has made HAR

F. Zhang et al. (Eds.): AIS&P 2024, LNCS 15399, pp. 1–11, 2025.
https://doi.org/10.1007/978-981-96-1148-5_1

applications more convenient and widespread. In everyday life, smartphones and wearable smart devices typically integrate multiple sensors. The real-time data collected by these devices can be analyzed using trained HAR models, allowing for precise analysis of the user's activity status. For instance, HAR technology can be embedded into wearable smart devices or smartphones to log personal activities, calculate daily calorie expenditure, and provide dietary and fitness recommendations. Additionally, in elderly care, HAR systems can monitor fall events and immediately issue alerts, thereby enabling swift assistance in emergencies and preventing potentially catastrophic injuries [3].

Suppose a medical research institution, Institution A, needs to train a HAR model for monitoring patients' post-operative daily activities. In that case, it would require a large amount of data to support the model's training. Due to significant variations in individuals' physical conditions, activity habits, and postures, training a high-quality HAR model necessitates collecting activity data from numerous diverse individuals. However, as awareness of data privacy and security continues to grow, the cost and complexity of acquiring such data have increased [4]. Therefore, federated learning [5,6] can be considered as a viable approach for model training. Institution A, acting as the server, can invite research institutions, enterprises, or individuals with relevant data to participate in joint model training by offering certain incentives. Within the federated learning framework, these participants are not required to disclose their local private data; instead, they upload the models trained on their local data to the server, thereby ensuring a high degree of data security. This approach also allows for a more diverse set of training data [7].

Although federated learning enables the training of models by uniting different participants over the internet without requiring data sharing, it introduces new risks. In this framework, the server coordinating the model training cannot access the training data of each participant; it can only view the models or model parameters uploaded by the participants. Consequently, if there are malicious participants in the system, they may attack the global model by modifying their local training data or poisoning their local models. After these harmful models are uploaded, the server cannot directly determine whether the models are harmful.

In this paper, we explore a feasible backdoor attack method in the training process of sensor-based HAR models within a Federated Learning environment. We further discuss how to mitigate the impact of backdoor attacks from malicious clients on the global model at the server side. When training and using HAR models, whether RNN or CNN models, it is necessary to preprocess the raw data collected by sensor devices, such as segmentation. Observations from publicly available datasets collected in free-living environments reveal that the raw sequential data gathered by sensors for HAR is typically lengthy and contains multiple types of activity data, with each time point corresponding to a specific activity label. To effectively process this raw sensor data, a common approach is to segment the long sequences into shorter time series using fixed-size time windows, and then assign activity labels to these segmented sequences. In free-living environments, different types of activities often alternate in the data, meaning a time window may contain data from multiple activity types. The usual practice is to assign the label of the most frequent activity type within that window. Based on this real-world scenario, we propose a method to disrupt the model during training through backdoor attacks, causing the model to predict a less frequent activity type during inference. If the input data during prediction contains features similar to the backdoor, the backdoor will be triggered, leading to the output of a preset result, even if this feature represents only a small portion of the data. Moreover, in HAR data, it is common for different individuals performing the same activity to generate similar data, making it easier for the backdoor to be triggered.

Experimental results show that our designed attack method can be effectively implemented in federated learning HAR training systems. Even without intentionally using backdoor data for training, the model can still predict backdoor-triggered test data as the preset activity type in some cases. Although the global model in a federated learning system is an aggregation of different client models, and the contribution of a malicious client to the model is relatively small, our method still achieves up to excellent attack success rates on public datasets.

2   Related Work

2.1   Human Activity Recognition

In sensitive research areas such as HAR-based healthcare, the scarcity of labeled data leads to poor model performance. To address this issue, [8] proposed the CapsLSTM model. Similar challenges also exist within the federated learning (FL) framework, where semi-supervised federated learning approaches have been adopted by [9,10] to mitigate these problems.

The heterogeneity of user data and the varying computational capabilities of devices in real-world scenarios pose higher demands on the effectiveness and accuracy of FL models. [11] addressed this by leveraging contrastive learning and adaptive control variables to handle biases between HAR clients. This approach reduces the representation gap between the global and local models, promoting the convergence of the global model.

Additionally, [12] introduced meta-learning into federated learning and proposed Meta-HAR, which significantly improved the personalized performance of HAR models. Despite individual differences in HAR data, the overall execution of actions remains similar. [13] utilized the similarity between different users' data for cluster-based learning, proposing ClusterFL and designing two novel mechanisms to enhance accuracy and reduce communication overhead.

In practical federated HAR applications, more challenges may arise. We must comprehensively consider the accuracy, fairness, robustness, and scalability of models. Moreover, the differences in performance and models of wearable devices used by different users may render traditional FL schemes ineffective for collaborative training. [14] proposed Hydra, which uses BranchyNet to design a mixed-size global model, allowing heterogeneous devices to train parts of the model suited to their computational capabilities. This approach clusters devices based on model similarity, reducing the impact of HAR data heterogeneity on model accuracy. The data collected by different clients may be of different modalities, and existing work on cross-modal federated human activity recognition has been insufficient. [15] proposed a cross-modal collaborative activity recognition network to overcome these deficiencies.

2.2 Backdoor Attack

Gu et al. [16] pointed out in the BadNets paper that when model training tasks are outsourced, attackers may maliciously implant backdoors into deep neural networks (DNNs). These backdoored models can be highly covert and harmful. There has been substantial research on backdoor attacks across various domains [17,18], with several defense methods proposed [19]. However, implementing a high attack success rate without sacrificing stealthiness is more challenging in sequential data due to its lower input dimensionality and degrees of freedom. Ding et al. [20,21] explored the vulnerabilities of sequential data to backdoor attacks. Similarly, Jiang et al. [22] identified a gap in backdoor attack research on sequential data and designed trigger generation methods capable of achieving high stealthiness and high attack success rates. Unlike centralized training methods, in federated learning, the data used by each participating client is only visible to that client, and clients only upload model parameters after completing local training. This scenario provides malicious clients with more opportunities to conduct poisoning attacks on data or models [23,24]. Backdoor attacks are also recently applied to many vertical domains such as machine unlearning performance evaluation [25,26].

3 Backdoor Attack on Federated HAR

During the process of training models using federated learning, the server is responsible for aggregating model parameters and does not have access to the data held by the participating clients. Consequently, if a malicious client exists within the system, it can modify its local training data by adding backdoor information, train the model, and then upload it to the server. In this scenario, the server cannot directly detect that the uploaded model has been compromised with a backdoor. As a result, when the server aggregates models from malicious clients, the global model becomes vulnerable to backdoor attacks. When backdoor features are present in the data, the model will output results predetermined by the attacker.

In previous studies on backdoor attacks involving time-series data, a common approach has been to train a trigger generator, which converts clean samples into backdoor samples. However, when attacking models trained on sensor data from hardware devices, potential attack methods can be considered from the hardware layer [27]. Therefore, in designing backdoor triggers for HAR data, we take into account potential security vulnerabilities that could be exploited by attackers based on the practical applications of HAR.

3.1   Signal Characteristic of HAR

In daily life, HAR is widely applied, particularly in wearable smart devices such as smartphones and smartwatches, and is extensively used for health monitoring. These devices use built-in sensor hardware (e.g., accelerometers and gyroscopes) to collect data and input it into the device's HAR model f_w to determine the user's current activity, where w represents the model parameters.

Common sensor-based HAR public datasets typically involve the collection of activity data from volunteers over a period. Some datasets collect data on individual activities separately, eliminating the need to label each data point in the raw data. However, many datasets capture volunteers' free activity data over time, requiring the labeling of each data point with the corresponding activity type.

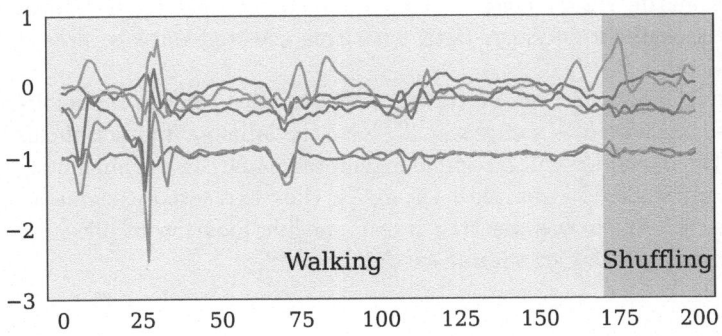

Fig. 1. A continuous segment of data selected from HAR70+ raw data using a time window of size 200. The light blue section (left) represents the activity type *Walking*, while the light red section (right) represents the activity type *Shuffling*. (Color figure online)

When training HAR models with CNN structures, a sliding window approach is typically used to divide the collected HAR raw data $\mathcal{X}$ into shorter time series segments $x \in \mathcal{X}$. If the data consists of HAR recordings from free activities over a period, where each data point in the raw data is labeled with an activity type, the segmented data should be labeled based on the proportion of the activity types it contains, usually assigning the label corresponding to the most frequent activity type in that segment. For example, the HAR70+ dataset comprises activity data collected from volunteers engaging in free activities over two hours. During model training, a time window of size 200 is used to segment the raw data. As shown in Fig. 1, the segmented data contains the most frequent activity type *Walking*, so it is labeled as *Walking*.

In real-world applications of HAR, users perform activities according to their habits, and the data collected by devices in everyday life is subject to more interference, with multiple activities often occurring simultaneously. When the

HAR model f_w is deployed on a device, the real-time activity data collected by the device is preprocessed and input into the model for prediction. After preprocessing, a segment of data x may contain data from two different activity types (the larger the time window used for segmentation, the more likely this situation occurs). A normally trained model f_w should predict x as the activity type with the higher proportion, $f_w(x) \rightarrow y$. However, if an attacker generates backdoor data, and the implanted backdoor information T is similar to the less frequent activity data, the model may predict x as the attacker's predetermined result, $f_w^*(x) \rightarrow y^*$.

3.2   Time-Series Alignment Based Backdoor Injection

To provide a simple illustration of generating backdoor data for a backdoor attack on a federated learning HAR system, this process involves selecting a sample from the target class as the source of the backdoor information and replacing a segment of clean data with this selected data to create backdoor data.

Specifically, let the set of target class samples on the malicious client be denoted as X_t, where all samples in X_t have the label y_t. The set of clean samples to which the backdoor will be added is denoted as X_o. In our initial experiments, we randomly selected a sample x_t from X_t, then extracted a segment from x_t as the trigger T, which was inserted into a specific location in all samples of X_o, resulting in the backdoor sample set X_b.

4   Experment Evaluation

4.1   Datasets

In this section, we provide a detailed overview of the datasets used in our experiments, including their sources, characteristics, and scales.

We conducted experimental validations on six publicly available HAR datasets. These datasets encompass HAR data collected under various environments, sensors, and age groups from different volunteers, making them naturally suitable for federated learning (FL) experimental settings. In our experiments, each subject within these datasets is treated as a participating client. The data collection methods across these six datasets include both free-form activities and individually recorded activities. Notably, the free-form activity datasets closely align with real-world HAR applications and serve as the foundation for our research on backdoor attacks in HAR systems.

Human Activity Recognition Using Smartphones (UCI HAR) [28]: This dataset comprises six different activities, including walking, sitting, standing, ascending stairs, descending stairs, and lying down.

Wireless Sensor Data Mining (WISDM) [29]: Activity data were collected by placing a smartphone in the pants pocket to simulate real-life daily scenarios, utilizing the phone's built-in sensors.

HAR70+ [30,31]: This unique dataset was collected from eighteen elderly individuals aged between 70 and 95 years, each wearing two accelerometer sensors. It includes recordings of eight different activities. The application of HAR among elderly populations holds significant importance, such as in elderly fall detection systems [32], which can help mitigate certain behavioral risks.

HARTH [33]: This dataset contains recordings from 22 participants who wore two 3-axis Axivity AX3 accelerometers for approximately two hours in free-living environments. A total of 12 different activities were recorded.

Motionsense [34]: This dataset comprises time-series data generated by accelerometer and gyroscope sensors. A total of 24 participants with varying genders, ages, weights, and heights performed six activities-descending stairs, ascending stairs, walking, jogging, sitting, and standing-across 15 trials under identical environmental conditions.

REALDISP [35]: This dataset consists of recordings of 33 fitness activities performed by 17 participants, captured using nine inertial sensor units.

4.2   Experiment Setup

In the validation experiments conducted on these six public datasets, we introduced a malicious client into the system. During each training round, the server randomly selected a subset of clients to participate in the training, with each selected client performing approximately 10 local training rounds. To evaluate the classification performance of the model, each client retained 20% of their data, denoted as D_{test}^i, as a test set. After the model training was completed, the global model's generalization ability was assessed on each client's test dataset D_{test}^i. To minimize the bias introduced by randomness, all experiments were repeated five times under identical settings, and the average of the five experimental results was reported, along with the deviation of each experiment from the mean.

For the malicious client C_{adv}, approximately 30% of its local training data (excluding target class data) was converted into backdoor data D_b^{adv} using the method described in Sect. 3. Additionally, 20% of the backdoor data was reserved as a test set D_b^{test} to evaluate whether the model was successfully attacked.

The primary metrics we focus on are:

– **Accuracy.** This metric measures the classification performance of the global model on the test data from all participating clients after training. The results presented in this paper represent the average Accuracy across all clients.
– **ASR (Attack Success Rate).** This metric measures the proportion of backdoor data that the model correctly classifies as the attacker's intended label, indicating the success rate of the backdoor trigger activation.

4.3   Backdoor Injection

In research on backdoor attacks against models, the desired outcome is to maintain the model's Accuracy on clean, normal samples while achieving a high ASR.

Therefore, we first conducted experimental validation of the backdoor insertion method proposed in Sect. 3. We compared two conditions: C_{adv} training its local model entirely on clean data (*clean*) and C_{adv} training its local model by adding backdoor data to the training set (*backdoor*). The experimental results are shown in Table 1.

The results in Table 1 indicate that the global model's Accuracy across the six datasets remains similar under both conditions, suggesting that the inclusion of backdoor data by the malicious client C_{adv} does not negatively impact the model's classification performance on clean samples. When C_{adv} added backdoor data during training, the global model achieved an ASR of up to 100% on five of the six datasets, with only the WISDM dataset yielding a lower ASR of 68.13%; the ASR exceeded 85% on all other datasets. Furthermore, even when C_{adv} trained its local model entirely on clean data, the average ASR on the backdoor test set D_b^{test} reached 21.36%, 3.12%, and 24.58% for the HAR70+, HARTH, and Motionsense datasets, respectively, indicating that backdoor attacks may succeed even without explicit training for backdoor features.

Table 1. Backdoor experiment result.

Dataset	Accuracy (%)		ASR (%)	
	clean	*backdoor*	*clean*	*backdoor*
UCI HAR	91.39	91.08	0	83.33
HAR70+	89.50	90.27	21.36	85.00
HARTH	89.19	89.97	3.12	96.53
Motionsense	91.89	92.64	24.58	100
REALDISP	92.56	92.44	0	96.88
WISDM	87.96	88.10	0	68.13

In federated learning-based model training attacks, the contribution of the malicious client to the global model is only partial in each round. The backdoor features learned by the malicious client's uploaded model are often diluted during the aggregation process. As shown in Fig. 2, the ASR fluctuates across training rounds for these three datasets. We marked the rounds in which the malicious client participated in training. Typically, the ASR of the model before aggregation is higher than after aggregation when the malicious client is involved in training. Additionally, if the malicious client is not selected in the initial rounds, the implanted backdoor can be quickly neutralized by other clients' models. However, despite the ASR fluctuating throughout the training process, the ASR tends to increase as the number of rounds involving the malicious client increases.

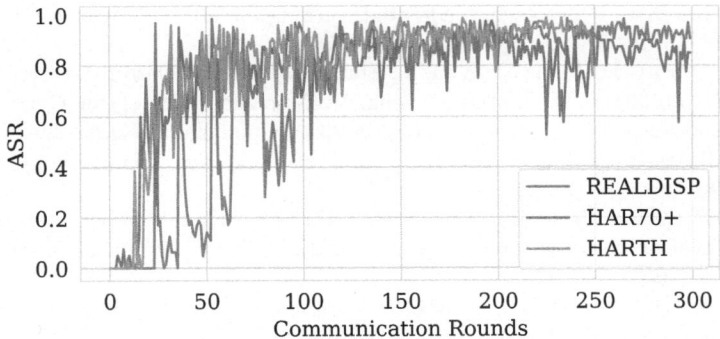

Fig. 2. ASR variation during training as federated communication rounds progress.

5   Conclusion

In this paper, we propose an effective backdoor attack method for time-series signals in human activity recognition (HAR) systems using federated learning, and we extensively validate this method on six public datasets, demonstrating its effectiveness. If a malicious client exists in a federated learning system and trains its local model using backdoor data, the global model will be impacted by the backdoor attack after the server aggregates the malicious client's model.

Acknowledgments. This work was supported by National Natural Science Foundation of China (No. 61802383), Research Project of Pazhou Lab for Excellent Young Scholars (No. PZL2021KF0024), Guangdong Regional Joint Fund Project (No. 2022A1515110157), University Research Project of Guangzhou Education Bureau (No. 2024312189), and Guangzhou Basic and Applied Basic Research Project (No. SL2024A03J00397).

References

1. Chen, K., Zhang, D., Mi, B.: Private data leakage in federated human activity recognition for wearable healthcare devices. CoRR **abs/2405.10979** (2024)
2. Chen, K., Zhang, D., Chai, Y., Zhang, W., Wang, S., Shen, J.: Federated unlearning for human activity recognition. CoRR **abs/2404.03659** (2024)
3. Kumar, P., Chauhan, S., Awasthi, L.K.: Human activity recognition (HAR) using deep learning: review, methodologies, progress and future research directions. Arch. Comput. Methods Eng. **31**(1), 179–219 (2024)
4. Lupión, M., Cruciani, F., Cleland, I., Nugent, C., Ortigosa, P.M.: Data augmentation for human activity recognition with generative adversarial networks. IEEE J. Biomed. Health Inform. **28**(4), 2350–2361 (2024)
5. McMahan, B., Moore, E., Ramage, D., Hampson, S., y Arcas, B.A.: Communication-efficient learning of deep networks from decentralized data. In: Artificial Intelligence and Statistics, pp. 1273–1282 (2017)

6. Chen, K.: Privacy preserving federated learning for full heterogeneity. ISA Trans. **141**, 73–83 (2023)
7. Zhang, X., Li, J., Zhang, J., Yan, J., Zhu, E., Chen, K.: Data reconstruction from gradient updates in federated learning. In: Machine Learning for Cyber Security - 4th International Conference, ML4CS 2022, Guangzhou, China, December 2-4, 2022, Proceedings, Part I. Lecture Notes in Computer Science, vol. 13655. pp. 586–596 (2022)
8. Khan, P., Kumar, Y., Kumar, S.: CapsLSTM-based human activity recognition for smart healthcare with scarce labeled data. IEEE Trans. Comput. Soc. Syst. **11**(1), 707–716 (2024)
9. Zhao, Y., Liu, H., Li, H., Barnaghi, P., Haddadi, H.: Semi-supervised federated learning for activity recognition (arXiv:2011.00851) (2021)
10. Yu, H., et al.: FedHAR: semi-supervised online learning for personalized federated human activity recognition. IEEE Trans. Mob. Comput. (2021)
11. Iwan, I., Yahya, B., Lee, S.L.: Federated Model Contrastive Learning with Adaptive Control Variates for Human Activity Recognition. Available at SSRN 4810020 (2024)
12. Li, C., Niu, D., Jiang, B., Zuo, X., Yang, J.: Meta-HAR: federated representation learning for human activity recognition. In: Proceedings of the Web Conference 2021, pp. 912–922 (2021)
13. Ouyang, X., Xie, Z., Zhou, J., Huang, J., Xing, G.: ClusterFL: a similarity-aware federated learning system for human activity recognition. In: Proceedings of the 19th Annual International Conference on Mobile Systems, Applications, and Services, pp. 54–66 (2021)
14. Wang, P., Ouyang, T., Wu, Q., Huang, Q., Gong, J., Chen, X.: Hydra: Hybrid-model federated learning for human activity recognition on heterogeneous devices. J. Syst. Architect. **147**, 103052 (2024)
15. Yang, X., Xiong, B., Huang, Y., Xu, C.: Cross-Modal Federated Human Activity Recognition. IEEE Trans. Pattern Anal. Mach. Intell. 1–18 (2024)
16. Gu, T., Dolan-Gavitt, B., Garg, S.: BadNets: Identifying vulnerabilities in the machine learning model supply chain. arXiv preprint arXiv:1708.06733 (2017)
17. Chou, S.Y., Chen, P.Y., Ho, T.Y.: VillanDiffusion: a unified backdoor attack framework for diffusion models. Adv. Neural Inf. Process. Syst. **36** (2024)
18. Wang, Y., Chen, K., Tan, Y., Huang, S., Ma, W., Li, Y.: Stealthy and flexible trojan in deep learning framework. IEEE Trans. Dependable Secur. Comput. **20**(3), 1789–1798 (2023)
19. Zhu, M., Wei, S., Zha, H., Wu, B.: Neural polarizer: a lightweight and effective backdoor defense via purifying poisoned features. Adv. Neural Inf. Process. Syst. **36** (2024)
20. Ding, D., et al.: Towards Backdoor Attack on Deep Learning based Time Series Classification, pp. 1274–1287 (2022)
21. Ding, D., Zhang, M., Feng, F., Huang, Y., Jiang, E., Yang, M.: Black-box adversarial attack on time series classification. Proc. AAAI Conf. Artif. Intell. **37**(6), 7358–7368 (2023)
22. Jiang, Y., Ma, X., Erfani, S.M., Bailey, J.: Backdoor Attacks on Time Series: A Generative Approach (2023)
23. Chen, K., Zhang, H., Feng, X., Zhang, X., Mi, B., Jin, Z.: Backdoor attacks against distributed swarm learning. ISA Trans. **141**, 59–72 (2023)
24. Bagdasaryan, E., Veit, A., Hua, Y., Estrin, D., Shmatikov, V.: How to backdoor federated learning. In: International Conference on Artificial Intelligence and Statistics, pp. 2938–2948 (2020)

25. Chen, K., Wang, Y., Huang, Y.: Lightweight machine unlearning in neural network. CoRR **abs/2111.05528** (2021)
26. Chen, K., Huang, Y., Wang, Y., Zhang, X., Mi, B., Wang, Y.: Privacy preserving machine unlearning for smart cities. Ann. Telecommun. **79**(1), 61–72 (2024)
27. Kim, H., et al.: A systematic study of physical sensor attack hardness. In: 2024 IEEE Symposium on Security and Privacy (SP), pp. 143–143 (2024)
28. Anguita, D., Ghio, A., Oneto, L., Parra, X., Reyes-Ortiz, J.L.: A public domain dataset for human activity recognition using smartphones. In: The European Symposium on Artificial Neural Networks (2013)
29. Kwapisz, J.R., Weiss, G.M., Moore, S.A.: Activity recognition using cell phone accelerometers. ACM SIGKDD Explor. Newsl **12**(2), 74–82 (2011)
30. Ustad, A., Logacjov, A., Trollebø, S.Ø., Thingstad, P., Vereijken, B., Bach, K., Maroni, N.S.: Validation of an activity type recognition model classifying daily physical behavior in older adults: The har70+ model. Sensors **23**(5), 2368 (2023)
31. Logacjov, A., Ustad, A.: HAR70+ (2023)
32. Chen, J., et al.: Digital twin empowered wireless healthcare monitoring for smart home. IEEE J. Sel. Areas Commun. (2023)
33. Logacjov, A., Bach, K., Kongsvold, A., Bårdstu, H.B., Mork, P.J.: HARTH: a human activity recognition dataset for machine learning. Sensors (Basel, Switzerland) **21** (2021)
34. Malekzadeh, M., Clegg, R.G., Cavallaro, A., Haddadi, H.: Mobile sensor data anonymization. In: Proceedings of the International Conference on Internet of Things Design and Implementation, pp. 49–58 (2019)
35. Baños, O., Damas, M., Pomares, H., Rojas, I., Tóth, M.A., Amft, O.: A benchmark dataset to evaluate sensor displacement in activity recognition. In: Proceedings of the 2012 ACM Conference on Ubiquitous Computing, pp. 1026–1035 (2012)

Fully Automated Generation Mechanism of Rootfs for Specified Operating Systems Under Linux

Guoquan Wang(✉), Jiaxuan Chen, and Chengfei Zheng

School of Computer Science, Qufu Normal University, Rizhao, China
wangguoquan03@foxmail.com

Abstract. As the file system that must be mounted before the operating system starts, the root file system (rootfs) is crucial to ensure the normal running of the operating system and is closely related to application security. However, the current generation method of rootfs is very complicated. At present, only manual generation and semi-automatic generation of rootfs are implemented in some operating systems. But there are some operating systems which do not even have the corresponding generating mode of it. Therefore, how to generate rootfs efficiently has become an urgent problem to be solved. In this paper, to reduce the complexity of the existing rootfs generating method and to generate rootfs that especially matches the OpenEuler operating system, we create an innovative fully automated generation mechanism for rootfs under the Linux operating system. By constructing a generic automatic generation template of rootfs, we realize the fully automatic generation of rootfs in the OpenEuler operating system and further extend it to other Linux operating systems, realizing the transformation of generating rootfs from semi-automatic to full-automatic. This mechanism greatly improves the efficiency of users in testing, developing, and deploying the rootfs of the operating system, at the same time effectively reduces the compatibility issues when it is deployed on different platforms. Currently, this generation mechanism has been adopted and put into use by the OpenEuler community. It not only provides technical support for the development of the OpenEuler operating system, but also provides a new solution for rootfs generation of other Linux operating systems. Hence, it has great practical value and application prospects.

Keywords: OpenEuler · Root File System · Full Automatic Generation · Application Security Testing

1 Introduction

The Linux operating system, as a model in the field of open-source and free software, has many notable advantages. Firstly, the openness of its source code allows developers and organizations around the world to freely access, modify, and use it, which greatly promotes the rapid growth of technological innovation

ⓒ The Author(s), under exclusive license to Springer Nature Singapore Pte Ltd. 2025
F. Zhang et al. (Eds.): AIS&P 2024, LNCS 15399, pp. 12–26, 2025.
https://doi.org/10.1007/978-981-96-1148-5_2

and software development [1]. Secondly, the stability and security of the Linux operating system have been widely recognized for its worldwide community support and strict open-source review mechanism. In addition, the Linux operating system also provides high customizability so that it can be used in different scenarios and satisfy a variety of needs.

Open-source operating systems have been a key research direction in the field of computer science and technology. OpenEuler, a server operating system created by Huawei in 2010 and open-sourced in 2019 [2,3], is a community edition of Linux. It inherits the advantages of Linux. By sharing capabilities with HarmonyOS, it can be used in various devices and scenarios. OpenEuler aims to improve the development environment and the performance of diverse computing power with all users and developers. However, its root file system (rootfs) cannot be generated by itself.

As a kind of file system, rootfs not only has the basic function of storing data, but is the first file system that must be mounted when the kernel is started [4]. It contains the critical catalogs and necessary files for the system to start and for other file systems to mount [5]. If it cannot be mounted from the specified device, other file systems cannot be properly loaded and the operating system cannot start as normal. Therefore, its stability and reliability have a decisive impact on the operation of the entire operating system [6].

At the present stage, the only way to generate a specified rootfs in a Linux operating system is to configure it manually, but this does not apply to the OpenEuler operating system. Therefore, how to implement the automatic generation of the specified rootfs in the OpenEuler operating system has become one of the key problems to be solved.

In this paper, to reduce the complexity of the existing rootfs generating method and to generate rootfs that matches the OpenEuler operating system, we create an innovative full-automatic generation mechanism of rootfs for specific operating systems. In addition, this study further extends the automatic generation mechanism to other Linux operating systems and realizes the automatic generation of rootfs across operating system platforms.

The innovations and features of this mechanism are as follows:

1. For the first time, the project implements the function of fully automatic generation of the rootfs for the OpenEuler operating system. With specific configurations and optimizations for OpenEuler, scripts can automatically handle the decompression of ISO images, the extraction of kernel and rootfs, the creation of VM images, and the generation of initramfs. The project has created rootfs for the OpenEuler operating system through automation tools from scratch. This breakthrough fills the gap in OpenEuler's fully automatic generation mechanism.
2. The project developed a set of generation mechanism for rootfs. Based on extracting virtual machine images, it can generate rootfs for various Linux operating systems. This mechanism uses the libguestfs library and the virt-install command set, which is suitable for a variety of Linux distributions (such as Ubuntu, CentOS, etc.), and supports users to quickly create rootfs

through automated scripts. Users can generate rootfs for different systems with simple commands. This is a technological leap for the generation of rootfs, which greatly lowers the technical threshold and improves the generation efficiency and stability.

3. The author writes complex automation scripts which can automatically identify the operating system version and ISO file path entered by the user, and automatically generate rootfs based on the type of operating system. This script not only performs well in generating rootfs for OpenEuler operating system, but can also be applied to other Linux operating systems, supporting user-defined configurations to meet the diverse needs of users. The cross-platform automated generation method greatly improves the efficiency of environment deployment and development testing, enables developers to quickly generate rootfs for different systems, and shortens the test cycle. This efficient, automated deployment tool saves developers a lot of time and reduces the errors that can be caused by manual work.

In conclusion, this mechanism not only offers technical support for the development of the OpenEuler operating system, but also provides a new solution for the generation of rootfs in other Linux operating systems. Hence, it has great practical value and application prospects.

2   Background

Currently, rootfs cannot be generated directly in Windows operating systems and MacOS. Usually, generation of it needs the help of a Linux operating system.

There are two common ways to generate rootfs in a Linux operating system:

One is to use the build tools to generate rootfs. One of the most popular build tools is Yocto Project [7]. It is an open-source project that aims to provide a set of tools and methods for generating rootfs for embedded Linux systems. By using the build engine called BitBake, Yocto Project builds a complete Linux distribution, which includes its kernel, rootfs, and various user space tools. Through the Yocto Project, users can generate customized rootfs according to specific requirements and can also edit the configuration file of the Yocto Project to add or remove software packages, adjust system settings and configuration files, etc., to meet their specific application requirements.

An alternative approach to customizing a root filesystem (rootfs) image is to utilize an existing one. In this method, users start with a pre-built rootfs image and modify it to suit their needs. The advantage of this approach is that it saves the time and effort associated with building a rootfs from scratch, which is particularly beneficial for applications that do not require extensive customization. Users can install, configure, and tailor software packages within the existing rootfs image to fulfill their specific requirements. This method is also well-suited for rapid prototyping and deployment. However, the process can be tedious due to the numerous steps involved in customization.

As an operating system that uses the Linux kernel, OpenEuler supports both of the aforementioned methods for generating a rootfs. Nonetheless, both approaches require users to manually create the rootfs, as there is no automated method available for generating it directly.

ISO is a storage format that contains a file system and a light disk with a complete data structure, so it can be used as the basis or part of a rootfs. Currently, the rootfs of the specified operating system cannot be generated directly from ISO on the Linux operating system. It mainly uses ISO to generate a virtual machine image (qcow2), and then extracts the kernel and rootfs through it. During this process, it is relatively difficult to generate a virtual machine image from an ISO. The following are the three generation methods commonly used in the field: (a) Use the qemu-img command to convert ISO into qcow2. Qemu-img cannot directly convert ISO to qcow2, but after a virtual machine (VM) is created by ISO, qcow2/raw can be the hard disk of the virtual machine. Therefore, we can create a qcow2/raw and then create a VM on this hard disk. After the system is installed, the conversion can be done. (b) Use the virt-install command to convert ISO into qcow2. That is, use the virt-install command tool and use ISO as the installation medium to satisfy the conversion requirements of image files in qcow2 format. (c) Use the virt-builder command to convert ISO into qcow2. That is, use the virt-builder command tool to copy and customize image files from an existing image template.

Creating a VM image by ISO requires manual configuration and is complicated. It is not only error-prone but also takes a long time. At present, the three commonly used methods of generating VM images from ISO have certain shortcomings and defects. Although the qemu-img command has the advantages of simple operation, good capability, and good cross-platform performance, it can only be used for simple conversion of hard disk image format to QCOW2. Due to its lack of some advanced features such as custom configuration and adding drivers or tools, it cannot be used for more complex customization and configuration.

As for the virt-install command, although it supports custom virtual machine configurations, interactive installations, and multiple installation sources, using it to install virtual machine images requires the user to manually specify various configuration parameters. This can be difficult for inexperienced users. Besides, it usually requires the user to manually input the required information during the installation process, and cannot realize the complete automatic installation. In addition, using the virt-install command to install the VM requires that users prepare the VM hard disk image file in advance, instead of directly creating the VM image file from ISO. Compared with the above two methods, the virt-builder command simplifies the installation process. It is compatible with many operating systems and supports automatic generation. But because it does not support custom configuration, and the installation process depends on a network connection, it still needs to be perfected.

As a Linux-based operating system, it is more difficult for OpenEuler to generate the rootfs of the specified operating system from ISO than on other Linux operating systems, especially in the process of converting ISO to qcow2.

However, it is difficult to generate rootfs through ISO directly at present, not to mention that there is no feasible method to automatically generate the rootfs of the operating system through ISO on the current OpenEuler operating system.

3 Preliminary Knowledge

This section covers the basic knowledge required to generate rootfs automatically in specified operating systems, including the features and advantages of the OCaml language, and an explanation of important files generated by Linux commands.

3.1 OCaml Language

OCaml is a powerful and multipurpose statically typed functional programming language [8]. It has a powerful static type system that can catch many common errors at compile time, which helps to improve the robustness and maintainability of code [9]. It supports functional programming features such as higher-order functions, anonymous functions, and recursion, making it possible to write concise and efficient functional style code. At the same time, OCaml has a built-in garbage collector that can automatically manage memory, reducing the complexity of manual memory management. Its type derivation system can automatically derive the type of most expressions, which reduces the need for explicit type declaration in code and improves the simplicity and readability of code. Therefore, it is suitable for software development and research work in various fields such as compiler and interpreter development, financial field, network services and distributed systems, scientific computing and numerical analysis.

3.2 Libguestfs and Virt-Install

Libguestfs is a collection of tools for managing virtual machine images [10]. It provides a set of tools and libraries for dealing with various virtual machine image formats. With libguestfs, users can perform operations on a VM image without starting a VM, including mounting a file system, reading/writing files, and installing software packages.

As for virt-install, it is a command-line tool for quickly creating and deploying virtual machines [11]. It provides a simple command line interface that allows users to specify virtual machine configuration parameters through command line options, including CPU, memory, storage, and so on. Virt-install supports automated installation of virtual machine operating systems and can be used in a variety of scenarios, including test, development, and production environments.

These two tools play an important role in virtualized environments. They help users manage and utilize virtualized resources. In the research described in this paper, the above tools will serve as the basis.

3.3   The General Process of Linux Command Generation

In the Linux environment, the process of generating commands is a systematic and automated project. Firstly, developers need to carefully write the source code in C, C++, or other programming languages. These codes form the core logic of the command. Then, by writing makefiles, we can define rules for compiling, linking, and building source code files, as well as dependencies between files. This converts source code into executable or library files.

Furthermore, developers use the Autoconf to generate a configuration file. This file is responsible for detecting the system environment and generating the final Makefile based on user-specified parameters and options. Next, by running the configure script, the system environment is checked, parameters are set, and Makefile is ready to be generated.

In some projects, you may also need to edit the Makefile.am file so that the Automake tool can generate the Makefile.in file based on the macros and rules defined in it. It is an intermediate product in the creation of Automake. Finally, by executing the make command and according to the rules in the Makefile, automatically complete the compilation, linking, and building, generating the final executable or library file. This marks the successful completion of the Linux command generation process. The series of steps not only improves development efficiency but also ensures the consistency and replicability of the building process.

3.4   Makefiles and Its Related Files

During the generation of Linux commands, it generally includes source code files, makefiles (which defines compilation and linking and building rules), configuration files (which configure and build parameters), Autoconf scripts (which generate configuration scripts), and Automake rules files (which generate Makefile.in files). Together, these things automate compilation, linking, and configuration. Among them, the compilation of source code is one of the key steps necessary to generate commands. The following is the details of this part.

3.4.1   Makefile, Makefile.in, and Makefile.am

In Linux command source code, there are three types of files related to build and compilation: Makefile, Makefile.in, and Makefile.am. These files are often used to automate the build process, especially in large projects. Here's how they work:

A Makefile is a text file that contains a set of rules and commands that describe how to compile, link, and build a software project. It defines the dependencies between files and how to convert source code files into executables or library files. Rules in a Makefile consist of targets, dependencies, and commands. When the file that the target depends on changes, the make command executes the corresponding commands according to the commands to update the target file.

The Makefile.in file is a template file that contains the variables and rules used to generate the final Makefile. The Makefile.am file is a part of the Autoconf and Automake toolchains and is used to describe how to convert source code files to Makefile.in files. In general, Makefile.am contains some specific macros and rules of Automake for generating Makefile.in files.

In summary, the Makefile is the actual file to build rules. It defines how to compile and link source code. The Makefile.in file is a template file for the Makefile. It contains the variables and rules used to generate the final Makefile. The Makefile.am file is a script file that describes how to generate Makefile.in file. It is usually automatically generated by the Automake. This layered structure makes the process of building the project more flexible and maintainable.

3.4.2   Autoconf and Automake

Autoconf and Automake are widely used automation tools in developing and building software projects.

Autoconf is used to generate the configuration script configure, which automatically configures the software source code based on the system environment [12] and the parameter user input to ensure that the software will compile and run correctly on different platforms.

Automake is used to generate makefiles [13]. It automatically generates makefiles based on the rules and file lists provided by the developer, simplifying the project build process and also ensuring that the project is compiled correctly on different operating systems.

These two tools are often used together, making the management and construction of software projects more convenient and reliable, while improving compatibility and portability across platforms.

4   Construction

To reduce the complexity of the existing rootfs generating method and to generate rootfs that matches the OpenEuler operating system or Linux operating systems, we create an innovative fully automated generation mechanism for rootfs under the Linux operating system. The construction of the generic automatic generation mechanism of rootfs is the key to the rapid and consistent deployment of the operating system. It greatly simplifies the configuration process, lowers technical barriers, and increases deployment efficiency. Through pre-defined mechanisms, users can ensure the consistency of the system and make the system adapt to the needs of different operating systems. In addition, the maintainability and ease of updating of the mechanism provide a solid foundation for long-term management and system security.

This article describes the Linux operating system which is under the designated rootfs automatic generation mechanism can generate rootfs of the following operating systems:

Linux operating systems: CentOS7 and CentOS8, all versions of OpenEuler, from RHEL3 to RHEL 8, Debian 6 and above, FreeBSD13 and FreeBSD14, Ubuntu 10.04 and above, Fedora 25 and above;

Windows operating systems: Windows 7, Windows Server 2008R2, Windows Server 2012, Windows Server 2012R2, Windows Server 2016 (Fig. 1).

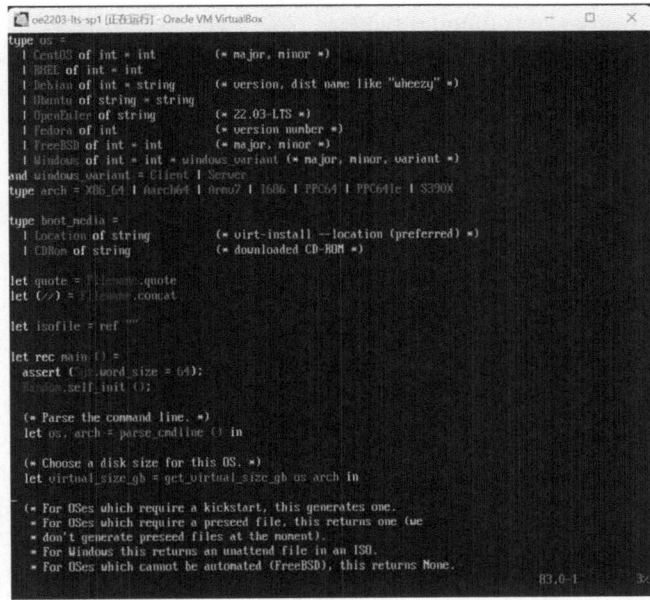

Fig. 1. The operating system version supported by the generated specified operating system mechanism under the Linux operating system

4.1 Fully Automatically Generate the Root File System for the Specified Operating Systems

1. Configure the generic generation templates for rootfs.
   - Install the packages libguestfs and virt-install.
   - Install the source package libguestfs.
   - Add a template for the specified operating system in the make-template.ml file under the builder/template/directory.
     (a) Open the make-template.ml file.
     (b) Use the function `get_virtual_size_gb` to get the virtual disk size `virtual_size_gb`.
     (c) Use the function `make_kickstart` to generate the kickstart file.
     (d) Use the function `make_boot_media` to get the boot media.
     (e) Use the command `tmpname` to create the domain name for the libvirt domain.
     (f) Use the function `make` to generate the kickstart file.
     (g) Use the command `output` to generate the output file name.
   - Generate an executable Makefile file
     (a) Write the command `virt-builder-template` to the Makefile.am file which is included in the `libguestfs-1.40.2/builder/` directory.
     (b) Return to the `libguestfs-1.40.2` directory.
     (c) Execute the command `automake`.
     (d) Execute the command `./configure`.

(e) Execute the command `make`.

Compile the target file for the generic rootfs generation template: execute the command `make` to build Makefile and the command `virt-builder-template`.

2. Generate a fully automated executable file of the generic rootfs generation template.
   - Go to the directory `libguestfs-1.40.2/builder`.
   - Execute the command `./virt-builder-template` (in the specified target operating system) to generate a mirror file.
   - Execute the command `xz -d` (in the specified target operating system) to extract the image file.
   - Execute the command `virt-builder -get-kernel` to extract the image kernel.
   - Execute the command `virt-tar-out -a` (in the specified target operating system) to extract rootfs.
   - Write a shell script that links to the above command and performs parameter parsing, environment validation, error handling, and disk space checking.

4.2 The Fully Automatic Generation for Rootfs for the Specified Operating System

1. Open the operating system on which you want to create the rootfs and enter the command line window.
2. Install virt-install, qemu-img, git software; use Git to clone the project source code to the operating system where the rootfs is to be generated. The operations are as follows:
   - Create a new directory and go to it.
   - Execute the command `git clone -single-branch -branch openEuler-22.03-LTS-Next` https://gitee.com/wangguoquan03/libguestfs.git.
   - Execute the command `git clone` https://gitee.com/openeuler/compass-ci/tree/master/lib/iso2rootfs. Go to the cloned libguestfs directory.
   - Unzip `libguestfs-1.40.2.tar.gz`.
   - Go to the directory `libguestfs-1.40.2`.
   - Execute the command $patch - p1 < ../oe - template.patch$.
   - Execute the command `automake`.
   - Execute the command `./configure`.
   - Execute the command `make`.
3. Automatically generate the rootfs for the specified operating system.
   - Return to the new directory.
   - Go to the directory `compass-ci/lib/iso2rootfs`.
   - Execute the command $./run - d < Dist > -r < Release > [-f] < ISO > [-p][/path/virt - x - dir/]$ to automatically generate the rootfs for the specified operating system.

– Among them, $-d <$ Dist $>$ is a required parameter, which is used to specify the target operating system name. $-r <$ Release $>$ is a required parameter, which is used to specify the Target Operating System version number. $[-f] <$ ISO $>$ is an optional parameter, which is used to specify the ISO file path. [-p] [/path/virt-x-dir/] is an optional parameter, which is used to build user-defined tool paths.

Figure 2 illustrates the logic flow chart of the tool.

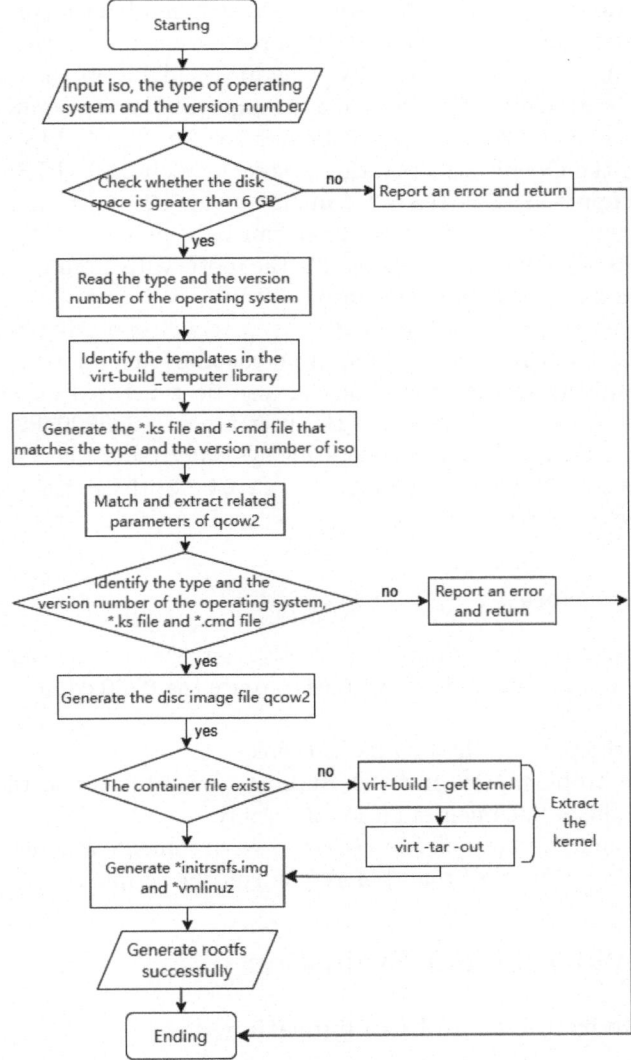

Fig. 2. Logic flow chart of the tool

4.3   The Use of the Fully Automatic Generation for Rootfs Under Specified Operating Systems

Because of the particularity of the OpenEuler operating system, there is no effective method to generate the rootfs of the specified operating system under the OpenEuler operating system. The study describes the generation process of it below.

1. Fully automated generation of the root file system (rootfs) for the specified operating systems under the OpenEuler operating system. Based on the above-mentioned generic automatic generation mechanism of rootfs, users only need to input iso2rootfs and its required parameters to generate rootfs of the operating system automatically. This tool is highly versatile and flexible, and the operation steps are the same for multiple Linux operating systems.
   For example, type ISO2ROOTFS -D OpenEuler -r 22.03-LTS to generate a rootfs with the OpenEuler operating system version 22.03-LTS.
   If the user inputs ISO2ROOTFS -D CentOS -r 8.5.2111, the rootfs with CentOS operating system version 8.5.2111 can be generated; if the user inputs ISO2ROOTFS -D ubuntu -r 22.04.03, the rootfs with Ubuntu operating system version 22.04.03 can be generated.
2. Fully automated generation of rootfs for specified operating systems under other Linux operating systems. Based on the above-mentioned methods, the rootfs of a given operating system can also be generated automatically on other Linux operating systems. If the user inputs ISO2ROOTFS -D Windows -r 7, the rootfs with Windows operating system version 7 is generated. If the user inputs ISO2ROOTFS -D ubuntu -r 22.04.03, the rootfs of the Ubuntu operating system version 22.04.03 is generated.

4.4   Points to Note

1. You may need to have administrator privileges when executing installation and compilation commands, so you need to use the SUDO command to elevate privileges.
2. Make sure the path to the ISO file is correct.
3. If you have problems compiling or running, you can refer to the README file or the official documentation for the study.
4. Make sure that you have all the necessary environments and dependencies in place as required by the study before you run the script.

5   Test Validation and Evaluation

5.1   Test Environment and Configuration

To fully evaluate the proposed fully automated rootfs generation mechanism, we tested it in a variety of hardware and software configurations. Figure 3 shows the configuration diagram of the host operating system. Figure 4 shows a partial log

```
[root@k8s-node1 sdb]# lscpu
Architecture:          x86 64
  CPU op-mode(s):        32-bit, 64-bit
  Address sizes:       39 bits physical, 48 bits virtual
  Byte Order:          Little Endian
CPU(s):                4
  On-line CPU(s) list: 0-3
Vendor ID:             GenuineIntel
  Model name:            Intel(R) Core(TM) i5-8265U CPU @ 1.60GHz
    CPU family:        6
    Model:             142
    Thread(s) per core: 1
    Core(s) per socket: 4
    Socket(s):         1
    Stepping:          12
    BogoMIPS:            3600.00
    Flags:               fpu vme de pse tsc msr pae mce cx8 apic sep mtrr pge mca cmov
pat pse36 clflush mmx
                       fxsr sse sse2 ht syscall nx rdtscp lm constant tsc rep good nopl
xtopology nonstop t
                       sc cpuid tsc known freq pni pclmulqdq vmx ssse3 cx16 pcid sse4 1
sse4 2 x2apic movbe
                       popcnt aes xsave avx rdrand hypervisor lahf lm abm 3dnowprefetch
invpcid single tpr
                       shadow vnmi flexpriority vpid fsgsbase bmi1 avx2 bmi2 invpcid
rdseed clflushopt md
                       clear flush l1d arch capabilities
Virtualization features:
  Virtualization:      VT-x
```

Fig. 3. The configuration of the host operating system

```
> > > > > > > > > > > > > > > >节选部分日志
[ 1557.984349][  T277] systemd-journald[277]: Received SIGTERM from PID 1
(systemd).
[ 1588.290085][   T1] SELinux:  policy capability network peer controls=1
[ 1588.290454][   T1] SELinux:  policy capability open perms=1
[ 1588.290650][   T1] SELinux:  policy capability extended socket class=1
[ 1588.291062][   T1] SELinux:  policy capability always check network=0
[ 1588.292035][   T1] SELinux:  policy capability cgroup seclabel=1
[ 1588.292175][   T1] SELinux:  policy capability nnp nosuid transition=1
[ 1588.292378][   T1] SELinux:  policy capability genfs seclabel symlinks=0
[ 1589.307298][  T24] audit: type=1403 audit(1698737094.276:2): auid=4294967295
ses=4294967295 lsm=selinux res=1
[ 1589.529659][   T1] systemd[1]: Successfully loaded SELinux policy in 4.863611s.
[ 1594.463954][   T1] systemd[1]: Relabelled /dev, /dev/shm, /run, /sys/fs/cgroup in
4.109914s.
[ 1595.035666][   T1] systemd[1]: systemd v249-16.oe2203 running in system mode
(+PAM +AUDIT +SELINUX -APPARMOR +IMA -SMACK +SECCOMP +GCRYPT
+GNUTLS -OPENSSL +ACL +BLKID -CURL -ELFUTILS -FIDO2 +IDN2 -IDN -IPTC
+KMOD -LIBCRYPTSETUP +LIBFDISK +PCRE2 -PWQUALITY +P11KIT -QRENCODE
+BZIP2 +LZ4 +XZ +ZLIB -ZSTD +XKBCOMMON +UTMP +SYSVINIT default-
hierarchy=legacy)
[ 1595.059069][   T1] systemd[1]: Detected virtualization qemu.
[ 1595.063543][   T1] systemd[1]: Detected architecture arm64.

Welcome to openEuler 22.03 LTS!
```

Fig. 4. Partial log of test results

of the test results. The test logs show that the tool can efficiently generate the rootfs of the specified operating systems.

The test environment includes, but is not limited to:

– Host operating system: Ubuntu 20.04 LTS, as a platform for development and testing.
– Target operating system: OpenEuler 22.03 LTS, as well as other Linux release versions as test subjects.
– Hardware configuration: Intel Xeon E5-2620 v4 CPU, 64GB of memory, 1 TB SSD disk, which ensure adequate computing and storage resources.
– Network environment: Gigabit Ethernet connection, average network latency of 5 ms, which ensure network stability.
– Software and tools: including but not limited to libguestfs, virt-install, qemu-img, XZ, gzip, tar, and performance testing tools such as top, htop, security checking tools such as SSH, ssh-keygen, and network testing tools such as Ping, Traceroute.

By comparing this mechanism with the traditional method of manually generating rootfs and the method of generating rootfs with some automatic tools, the advantages of this mechanism can be found (Table 1):

Table 1. Comparison of Rootfs Generation Methods

Property	This Mechanism	Traditional Manual Method(Extracting after the installation)	Partially Automated Tools(Use the instruction virt-bulider)
Generation method	Fully automatic	Manual	Semi-automatic
Compatibility	Cross-platform	Single platform	Multi-platform, but not supported OpenEuler
User interface	Command line interface	No	Command line interface
Deployment efficiency	High	Low	Middle
OpenEuler support	Yes	No	No
Expandability	High	Low	Middle
Customizability	High	Low	Middle
Error handling and reply	High	Low	Low
Promotion cost	Low	High	Middle
Ease of use	High	Low	Middle
Learning cost	Low	Middle	Middle
Stability	High	Low	Middle

During the compilation of the project, the code reserves the '-fPIE' compilation option. This is a common security measure while compiling to generate a Position Independent Executable(PIE). PIE makes it possible for executable to be randomly loaded in memory, strengthening the protection against buffer overflows and other memory attacks. By applying '-fPIE', the use of fixed memory addresses is avoided. It makes it impossible for attackers to easily predict the exact location of the program in memory, thus reducing the risk of being exploited for code injection attacks.

The code ensures the validity of the input parameters by parsing and verifying the parameters entered by the users. For example, if the 'getopts' command is used in the script to parse the input parameters and check the format and range of the individual parameters, when the user provides an invalid parameter, the script will return an error code and exit, preventing the program from continuing to run. This parameter verification mechanism can effectively prevent a variety of common attack vectors by strictly checking the format and validity of parameters entered by users. By verifying the type and range of parameters, the mechanism ensures that the script only accepts input that conforms to the expected format, preventing attackers from manipulating script by inputting illegally, such as inputting special characters or paths in the command lines. It not only prevents attackers from accessing and modifying sensitive files or directories, but also reduces script crashes or abnormal behaviors caused by accidental misinput, and guarantees the security and stability of script execution.

5.2 Evaluation

The results of the project evaluation show that the automation tools of this project are excellent in many aspects. In terms of functionality, the tool successfully generates the expected rootfs and is able to run on a variety of Linux operating systems such as CentOS, Ubuntu, and OpenEuler. Performance tests showed that the tool completes its task within a reasonable time. In terms of stability, the tool works well on different hardware platforms and network environments without major errors. The tool has a friendly interface, easy operation, high degree of automation and high degree of user satisfaction. The security test shows that the generated rootfs has no serious security holes, and the tool has good protection ability. After the security testing, the tool shows the expected execution and final result as shown in Fig. 2. The generated rootfs does not have serious security vulnerabilities and has good protection capabilities. In conclusion, the project achieves the expected goals, and in the functionality, performance, stability, user experience, and security have achieved significant results.

6 Conclusion

Rootfs, as the file system that must be mounted before the operating system starts, is crucial to ensure the normal running of the operating system. However, the current generation method of rootfs is very complicated. At present, only manual generation and semi-automatic generation of rootfs are implemented in some operating systems. But there are some operating systems which do not even have the corresponding generating mode of it. This paper proposes a fully automated generation mechanism of rootfs to simplify and accelerate the generation process of rootfs for OpenEuler and other Linux operating systems to ensure application security.

It develops a set of automated tools and processes that not only improve the efficiency of rootfs generation but also ensure the consistency and reusability of the generation environment. With these tools, developers can focus more on innovation and application development than on configuration and maintenance of the generation environment.

In the future, we plan to extend this mechanism to more operating systems and application scenarios. At the same time, we will also pay close attention to the feedback from the community to optimize the functionality of the tools and enhance the experience of users. In addition, security and performance will continue to be the focus of our efforts to ensure that the generated rootfs are secure and efficient.

References

1. Kroah-Hartman, G., Corbet, J., McPherson, A.: Linux Kernel Development. Addison-Wesley Professional, Boston, MA, USA (2010)
2. Zhou, M., Xinwei, H., Xiong, W.: openEuler: advancing a hardware and software application ecosystem. IEEE Softw. **39**(2), 101–105 (2022)
3. OpenEuler. Openeuler documentation. https://openeuler.org/en/. Accessed 15 May 2024
4. Malallah, H., et al.: A comprehensive study of kernel (issues and concepts) in different operating systems. Asian J. Res. Comput. Sci. **8**(3), 16–31 (2021)
5. Peng, L., Wang, Q.: Automated root file system generation for linux-based operating systems. In: 2019 IEEE International Conference on Systems, Man and Cybernetics (SMC), pp. 1322–1327 (2019)
6. Nemes, E.: Security aspects of root file systems in linux. J. Comput. Secur. **25**(4), 367–389 (2017)
7. Yocto Project. Yocto project documentation. https://www.yoctoproject.org/docs/. Accessed 15 May 2024
8. OCaml. Ocaml documentation. https://ocaml.org/docs/. Accessed 15 May 2024
9. Elliott, C., Hickey, T.: Programming with dependent types in OCaml. ACM SIGPLAN Notices **44**(9), 15–24 (2009)
10. Libguestfs. Libguestfs documentation. https://libguestfs.org/. Accessed 15 May 2024
11. Virt-install. Virt-install documentation. https://virt-manager.org/. Accessed 15 May 2024
12. Autoconf. Autoconf documentation. https://www.gnu.org/software/autoconf/. Accessed 15 May 2024
13. Automake. Automake documentation. https://www.gnu.org/software/automake/. Accessed 15 May 2024

Anti-side-channel Attack Mechanisms in Blockchain Payment Channels

Yijia Wang, Xinqi Dong, Jingxin Wang, Xu Zhang, Lianjin He, Mingyue Xu, and Yilei Wang[✉]

School of Computer Science, Qufu Normal University, Rizhao 276827, China
wang_yilei2019@qfnu.edu.cn

Abstract. Blockchain systems face increasing security threats, particularly the risk of side-channel attacks. In blockchain payment channels, the signing process of transactions often leaks a lot of sensitive information, making the system an easy target for attackers to launch side-channel attacks. Attackers can infer private data such as private key information by capturing physical information such as time during the signing process, which in turn compromises the security of the entire payment system. To effectively address this challenge, this paper proposes a novel signature algorithm specifically designed to defend against side-channel attacks. Specifically, a random delay is introduced into the signature algorithm as a random noise, which disturbs the side-channel information of the signature process, making it impossible for the attacker to infer the private data from the side-channel information in the channel. Through experimental verification, the signature algorithm proposed in this paper has a significant advantage in resisting side-channel attacks, with an improvement of 89.52% compared to the current payment channel system scheme.

Keywords: Blockchain · Side channel attack · Payment channel · Privacy security

1 Introduction

Blockchain technology originated from the peer-to-peer electronic transaction system proposed by Satoshi Nakamoto in 2008 [1], followed by the birth and rise of bitcoin. With the continuous advancement of the technology, blockchain has been widely used in many fields, such as finance [2], healthcare [3] and the Internet of Things (IoT) [4], and many different blockchain systems have been formed. However, these systems usually face the problem of data silos, which makes it difficult to circulate and share data between chains, limiting the scalability and interoperability of blockchain technology.

To address the scalability and interoperability issues of blockchain systems, virtual payment channels [5] have gained widespread attention as an efficient and secure cross-chain payment method. This technology allows users to complete cross-chain transactions without directly transferring on-chain assets by

F. Zhang et al. (Eds.): AIS&P 2024, LNCS 15399, pp. 27–34, 2025.
https://doi.org/10.1007/978-981-96-1148-5_3

establishing virtual channels between different blockchains. The virtual payment channel not only reduces transaction costs, but also significantly improves transaction speed and efficiency. The core idea is to lock the funds on the original blockchain and reflect the transaction results to the target blockchain through a secure and trustless mechanism, thus achieving cross-chain value transfer. The creation of cross-chain virtual payment channels [7] further enhances the interoperability of blockchains, allowing transactions to be transferred between different blockchains.

However, despite the significant advantages of cross-chain virtual payment channels in improving the efficiency of cross-chain transactions, the issues of privacy leakage and security should not be ignored. In cross-chain transactions, the transaction amount and status information may be exposed, resulting in privacy leakage. In addition, cross-chain virtual payment channels are also vulnerable to attacks, especially side channel attacks [8], which threaten the security of cross-chain transactions by analyzing the physical characteristics of the system to obtain sensitive information.

Therefore, although cross-chain virtual payment channels provide an efficient solution for cross-chain transactions, their privacy and security issues remain challenging. To address these issues, many technical means have been proposed to enhance privacy protection. This paper focuses on side channel attacks and proposes methods to address privacy issues in cross-chain virtual payment channels, aiming to improve the security and reliability of this technology.

1.1   Related Work

The decentralized nature of blockchain brings about scalability issues, and cross-chain technology has become the key to solving this problem. Payment channels improve transaction efficiency by pre-depositing funds on the blockchain. The most representative of these technologies is the Lightning Network proposed by Poon [9]. In order to further improve transaction throughput, Malavolta [10] proposed anonymous multi-hop lock technology in 2018, using encryption technology to protect user privacy. However, this technology requires intermediate nodes to be online all the time and is vulnerable to Domino attacks [14], which destroy the underlying payment channel.

To solve this problem, the Donnor protocol [14] constructed a virtual payment channel that only requires one user to provide funds, reducing dependence on the underlying payment channel and resisting Domino attacks. Cross-chain virtual payment channels [7] further improve this solution, requiring only intermediate nodes to participate in the opening, closing, and unloading operations of the payment channel, rather than being continuously online. However, the signature algorithm in the transaction process may be at risk of side channel attacks, threatening the security of the transaction.

However, the signature algorithms can be vulnerable to side channel attacks during the transaction process. These attacks can compromise the security of the transaction. Attackers can infer sensitive data [13] such as private keys and transaction behavior [6] in the blockchain system by monitoring the power consumption, latency, time and other information of nodes. For example, Rivest Shamir Adleman (RSA) and Elliptic Curve Digital Signature Algorithm (ECDSA) can be vulnerable to time analysis [11]. Attackers can target and attack specific miners or participants by detecting differences in the execution time of specific nodes [12].

1.2 Contribution

To address the above challenges, this paper proposes a generalized signature algorithm to resist side-channel attacks during cross-chain transactions. The algorithm reduces the possibility of the attacker to infer the transaction amount by using the state information such as time during the transaction process by introducing random delay as random noise, thus effectively protecting the transaction privacy. In order to verify the feasibility of resistance to side channel attack, this paper implements the signature algorithm in Python environment and uses the time ratio as the evaluation basis. The experimental results show that the algorithm significantly improves the system's resistance to side channel attacks.

1.3 Roadmap

The rest of this paper is organized as follows: Sect. 2 reviews the background of virtual payment channels and side channel attacks. Section 3 describes in detail the proposed signature algorithm, which is resistant to side channel attacks, and analyses the security of the algorithm. The complexity and scalability of the algorithm are also analyzed. In Sect. 4, we evaluate the efficacy of the algorithm. Finally, we conclude this paper in Sect. 5.

2 Preliminaries

2.1 Virtual Payment Channels

Virtual channel technology [5] significantly improves the scalability and transaction throughput of blockchain payments. The technology allows two users to establish a direct channel through a third-party intermediary. For example, there is channel β_1 between Alice and Bob, and there is channel β_2 between Bob and Caron. If Alice and Caron want to build a virtual channel γ, they can use Bob to build it using channels β_1 and β_2. The specific construction process of the virtual channel is shown in Fig. 1.

2.2 Side Channel Attacks

Side channel attacks are attacks based on additional information collected from the implementation of a computer protocol or algorithm. By obtaining this side channel information, the attacker is able to recover sensitive data [15]. Even if the

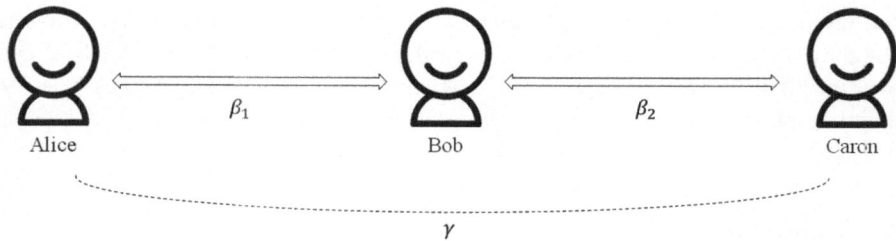

Fig. 1. Virtual payment channel.

data is encrypted, the attacker can still infer sensitive information by analyzing these physical properties. In cross-chain transactions, the attacker may be able to infer the transaction amount by analyzing the side channel information (such as time and power consumption) during the transaction, thereby leaking privacy.

3 Anti-side-channel Attack Mechanism

3.1 Algorithm Description

In order to enhance the system's ability to resist side channel attacks, this paper introduces randomization (*e.g.*, random delay, random noise) into the computation process to disrupt the side channel information and increase the difficulty for the attacker to conduct effective analysis. In this paper, we propose a signature algorithm against side channel attacks for the signature operation of user identity when constructing the Tx_f of the ledger channel in Bitcoin transactions.

Specifically, in this signature algorithm sign_input, the ability of the payment system to resist side channel attacks is improved by adding a random noise when dealing with identity-related keys and addresses. This random noise is modelled by an algorithm that adds a random delay, which is set in this paper in the range of [0,0.005] (in milliseconds). The core function of the algorithm is shown in Algorithm 1. In this paper, we evaluate the ability of anti-side-channel attacks by setting up an evaluation algorithm that uses the ratio of the difference between the maximum and minimum times to the average time as a performance evaluation criterion.

3.2 Security Analysis of the Algorithm

The random delay introduced is between [0, 0.005] seconds, which is a relatively short period of time, but sufficient to break the exact relationship between execution time and cryptographic operations to some extent. In side channel attacks, attackers typically infer the behavior of operations involving private keys or other private information by monitoring the execution time of the device. By adding random delays to the key signature steps, this randomization process effectively makes it more difficult for an attacker to collect sufficiently accurate data. The

---

Algorithm 1 Core part of signature algorithms against side channel attacks

---

1: **function** INIT(*self*, *sk*)
2:     *self.sk* = Create a private key object using private key *sk*
3:     *self.pk* = Get the public key from the private key object
4:     *self.addr* = Get the address from the public key object
5:     *self.p2pkh* = Use the address to create a script to pay to the public key hash
6:     call *add_random_delay*()
7: **end function**
8: **function** SIGN_INPUT(*self*, *tx*, *input_index*, *script*)
9:     *hash_to_sign* = Get the hash value corresponding to the input indexed as *input_index* in transaction *tx*
10:     *signature* = Sign the hash value *hash_to_sign* using the private key *self.sk*
11:     call *add_random_delay*()
12:     **return** *signature*
13: **end function**
14: **function** ADD_RANDOM_DELAY(*self*)
15:     *delay* = Generate a random delay between 0 and 0.005 seconds
16:     Pause execution for *delay* seconds
17: **end function**

---

execution time of the algorithm is no longer directly correlated to a specific operation, making it harder for an attacker to perform statistical analysis based on timing discrepancies, thereby increasing the blockchain system's resistance to side channel attacks.

3.3   Complexity and Scalability Analysis of the Algorithm

Complexity. From the point of view of the complexity of the algorithm, the addition of random delays does not significantly increase the computational complexity of the algorithm; the delays are implemented by systematic waiting times. This means that although an additional time delay is introduced, the core part of the signature algorithm still retains its original computational complexity. At the same time, the randomly delayed execution time does not affect the validity and correctness of the transaction.

Scalability. In a multi-node, multi-signature scenario, the signature algorithm can be scaled up and applied to more complex systems. The random delay introduced by each node when generating signatures is independent, which can effectively prevent attackers from analyzing the relationship between multiple signature operations through time correlation. At the same time, the algorithm can be combined with other side channel protection measures to further enhance the overall attack resistance of the system.

4   Performance Evaluation

In this section, this paper evaluates the proposed signature algorithm's ability to withstand side channel attacks and its performance in enhancing the privacy protection of Bitcoin data transactions. In order to measure the algorithm's ability to withstand side channel attacks, the evaluation criterion "time imbalance" is used, as shown in Equation (1). Where T_{max} denotes the maximum time, T_{min} denotes the minimum time and T_{avg} denotes the average time. The time imbalance is obtained by calculating the difference between the maximum time and the minimum time in the signature process and analyzing the ratio with the average signature time. The size of this ratio directly reflects the stability of the algorithm's execution time. The larger the ratio, the greater the time variation of the algorithm, which makes it easier for side channel attackers to obtain sensitive information through time analysis and increases the risk of successful attacks.

$$Time\ Imbalance = \frac{T_{max} - T_{min}}{T_{avg}} \tag{1}$$

The signature algorithm proposed in this paper effectively reduces the unevenness of the signature time, thus enhancing the protection against side channel attacks. In order to visualize this improvement effect, Fig. 2 gives the comparison results of different algorithms in terms of time imbalance.

The green line represents the signature algorithm against side channel attacks proposed in this paper, which exhibits significant advantages against side chan-

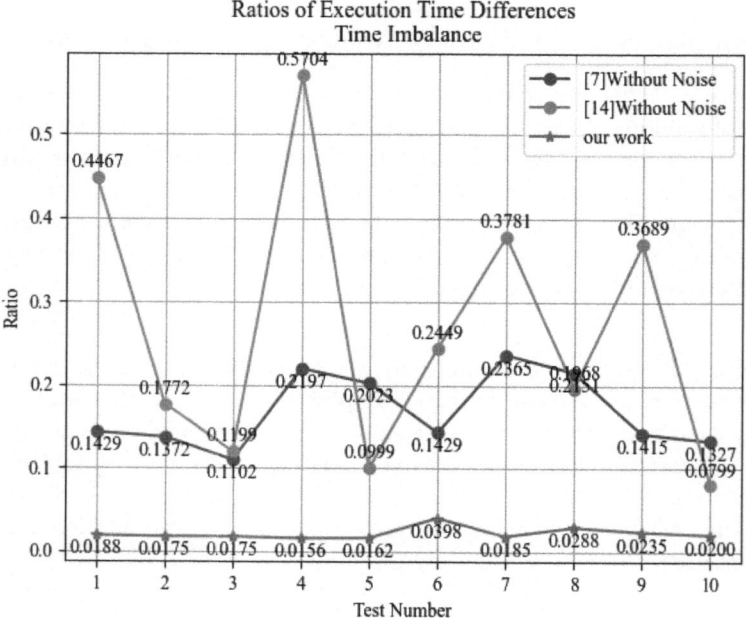

Fig. 2. Comparison of side channel attack capabilities.

nel attacks. Specifically, the algorithm significantly reduces the rate of temporal imbalance, making the system more robust against side channel attacks. Compared to the algorithms in [7] and [14], the method proposed in this paper is 89.52% more resilient to side channel attacks. This indicates that the new algorithm has significant improvements in security and attack resistance.

5 Conclusion

Cross-chain virtual payment channels are an effective solution to the scalability problem of cross-chain transactions of cryptocurrencies and have been put into practical use. Although there are many existing solutions that attempt to solve the scalability challenges in cross-chain transactions, they are still vulnerable to side channel attacks and data privacy leaks. To this end, this paper proposes a signature algorithm that is resistant to side channel attacks and aims to ensure the security of transactions. However, there are limitations to this work. In large systems, the signature operation of each transaction adds random delays. If the system has a very large number of concurrent transactions, these small delays can add up and have some impact on the overall throughput of the system. Therefore, future research can continue to develop more lightweight algorithms while maintaining a balance between performance and security.

References

1. Nakamoto, S.: Bitcoin: A peer-to-peer electronic cash system (2008)
2. Dong, C., Huang, Q., Fang, D.: Channel selection and pricing strategy with supply chain finance and blockchain. Int. J. Prod. Econ. **265**, 109006 (2023)
3. De Aguiar, E.J., Faiçal, B.S., Krishnamachari, B., Ueyama, J.: A survey of blockchain-based strategies for healthcare. ACM Comput. Surv. **53**(2), 1–27 (2020)
4. Koshy, P., Babu, S., Manoj, B.S.: Sliding window blockchain architecture for Internet of Things. IEEE Internet Things J. **7**(4), 3338–3348 (2020)
5. Dziembowski, S., Eckey, L., Faust, S., Malinowski, D.: Perun: virtual payment hubs over cryptocurrencies. In: 2019 IEEE Symposium on Security and Privacy, pp. 106–123. (2019)
6. Mangard, S., Oswald, E., Popp, T.: Power Analysis Attacks: Revealing the Secrets of Smart Cards. Springer Science & Business Media **31** (2007)
7. Jia, X., Yu, Z., Shao, J., Lu, R., Wei, G., Liu, Z.: Cross-chain virtual payment channels. IEEE Trans. Inf. Forensics Secur. **18**, 3401–3413 (2023)
8. Standaert, F. X.: Introduction to side-channel attacks. In: Verbauwhede, I. (eds), pp. 27–42. Springer, Boston, MA (2010). https://doi.org/10.1007/978-0-387-71829-3_2
9. Poon, J., Dryja, T.: The bitcoin lightning network: Scalable off-chain instant payments (2016)
10. Malavolta, G., Moreno-Sanchez, P., Schneidewind, C., Kate, A., Maffei, M.: Anonymous multi-hop locks for blockchain scalability and interoperability. In: Network and Distributed System Security Symposium (2019)
11. Kocher, P.: Differential power analysis. In: Proceeding Advances in Cryptology (1999)

12. Conti, M., Kumar, E.S., Lal, C., Ruj, S.: A survey on security and privacy issues of bitcoin. IEEE Commun. Surv. Tutorials **20**(4), 3416–3452 (2018)
13. Author, F.: SoK: communication across distributed ledgers. In: Borisov, N., Diaz, C. (eds) FC 2021, LNCS, vol. 12675, pp. 3–36. Springer, Berlin, Heidelberg (2021). https://doi.org/10.1007/978-3-662-64331-0_1
14. Aumayr, L., Moreno-Sanchez, P., Kate, A., Maffei, M.: Breaking and fixing virtual channels: Domino attack and donner. In: Network and Distributed System Security Symposium (2023)
15. Joy Persial, G., Prabhu, M., Shanmugalakshmi, R.: Side channel attack-survey. Int. J. Adv. Sci. Res. Rev **1**(4), 54–57 (2011)

F2L: A Lightweight Focus Layer Against Backdoor Attack in Federated Learning

Peiyao Niu, Lumin Zhou, Yan Lv, Peilin Li, Yue Wang, and Tao Li[✉]

School of Computer Science, Qufu Normal University, Rizhao 276827, China
litao_2019@qfnu.edu.cn

Abstract. Federated Learning (FL) is a distributed machine learning technique that enables clients to train Deep Neural Networks (DNNs) collaboratively without centralized data storage. However, this distributed nature introduces security challenges, making FL vulnerable to backdoor attacks. In these attacks, an attacker injects triggers into local models and tampers with updates during aggregation, causing incorrect predictions for specific inputs. As backdoor attacks evolve to become more covert, detecting these stealthy backdoors has become a critical issue. In this paper, we design the Lightweight Focus Layer (F2L), an adjustable layer inserted into server aggregation rounds. F2L detects subtle variations in backdoor attack parameters and fine-tunes them to minimize their impact. Our approach can be combined with existing defenses, significantly enhancing the robustness of FL models. Testing on three public datasets showed that F2L maintained the main task's accuracy while significantly reducing backdoor attack effectiveness.

Keywords: Federated Learning · Lightweight Focus Layer · Defense against backdoor attacks

1 Introduction

Federated Learning (FL) [1] enables multiple clients to collaboratively train deep neural networks (DNNs) on their private data. The server distributes a global model, and clients train it locally, uploading updates back to the server. FL addresses data privacy and security concerns by ensuring the server remains unaware of client data, unlike traditional centralized machine learning (ML).

However, FL also brings new privacy and security challenges [2–4]. Since client data and model training are black boxes to the server, FL is vulnerable to backdoor attacks [5]. During training, an attacker can inject triggers, causing the model to behave maliciously under specific conditions while functioning normally otherwise. As backdoor triggers become more covert, recent work like IBA (Irreversible Backdoor Attack) [6] shows how attackers enhance stealth. In IBA, imperceptible perturbations are added to clean images, creating minimal differences in latent space heatmaps. To maintain the attack's persistence, IBA

F. Zhang et al. (Eds.): AIS&P 2024, LNCS 15399, pp. 35–43, 2025.
https://doi.org/10.1007/978-981-96-1148-5_4

targets parameters less likely to be updated during main task learning and leverages previous global model parameters to guide current updates, ensuring the backdoor's lasting impact.

Current defense methods against backdoor attacks in FL [7–11] are ineffective against IBA, as they focus on pruning, fine-tuning, or adding noise to frequently updated parameters. IBA, however, targets less frequently updated parameters, bypassing these defenses. To address this, we propose a Lightweight Focus Layer (F2L), which fine-tunes parameters with update frequencies below a threshold B using a two-step method: (1) Project these parameters onto the direction of the previous global model, and (2) Use the Krum method to select and average parameters, forming new ones. This significantly reduces the impact of covert backdoor attacks. Our method, when combined with existing defenses, outperformed state-of-the-art (SOTA) methods on MNIST, CIFAR-10, and TinyImageNet in defending against backdoor attacks. Our contributions are summarized below:

- We propose F2L, a backdoor defense method that targets federated aggregation rounds with fewer updated parameters, which are more susceptible to attacks, and fine-tunes these parameters to significantly reduce the backdoor's impact.
- Our method can be integrated into existing defenses as a module, and experiments show that it helps reduce backdoor attack accuracy.
- We evaluated our method as a sub-module added to existing models on MNIST, CIFAR-10, and TinyImageNet datasets, and it outperformed SOTA defenses in terms of performance.

2   Related Work

Backdoor attack strategies: Backdoor attacks in FL have been widely studied and are evolving to be more potent, covert, and persistent [12,13]. IBA [6], proposed by Dung Thuy Nguyen et al., targets parameters updated less frequently during global aggregation. Using Poisoning-space and Poisoning-dimension methods, IBA sustains high attack performance over time and effectively bypasses existing defenses, posing a significant threat to the privacy of Federated Learning systems.

Backdoor defense strategies: The Krum algorithm selects the global model update by minimizing the sum of Euclidean distances between each client's update and others [7]. Norm Difference Clipping (NDC) reduces the impact of backdoor attacks by clipping the L2 norm of updates to a predefined range, mitigating the effect of large updates [8]. RFA (Robust Federated Aggregation) uses the geometric median, computed efficiently with a Weiszfeld-type algorithm, to maintain privacy and robustness without disclosing individual clients' contributions [14]. FLAME [15] integrates adversarial mitigation with model ensemble techniques, detecting backdoor attacks by analyzing client update similarities and using ensemble methods to filter out malicious updates, thus enhancing global model robustness.

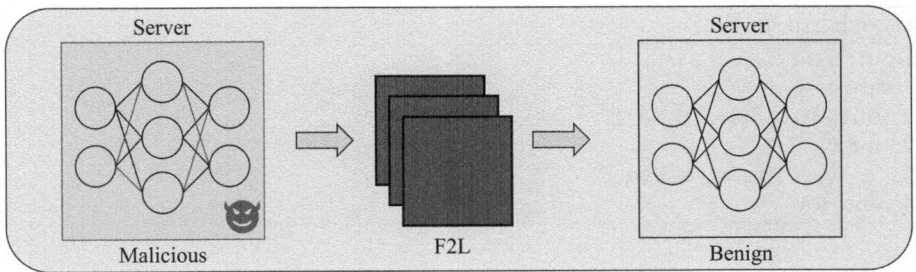

Fig. 1. Overview of F2L.

3   Method

F2L. Our goal: For IBA, we designed a lightweight backdoor focus layer that fine-tunes parameters updated less frequently during global model aggregation rounds, thereby mitigating the impact of backdoor attacks. Specifically, we define a parameter change boundary B, Let w_j ($w_j < B$, $j \in b$) be the parameters updated less frequently than B, and b represents the number of parameters with update frequencies below B, the selected parameters are fine-tuned using the Eq. 1

$$\hat{w}_j = Avg(w_j + K(w_i)), \tag{1}$$

where $Avg(\cdot)$ denotes the averaging function, $K(\cdot)$ represents the Krum method, and w_i refers to the parameters in n that are not in b and exclude j, The parameter $\hat{w}_j$ is used as the new parameter in the current round of server aggregation, thereby mitigating the impact of the backdoor attack.

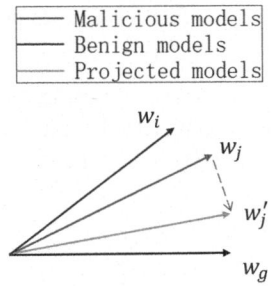

Fig. 2. Gradient Projection.

Algorithm 1. F2L

Input: B, i, j, n, T, $\mathbf{w}$
Output: $\hat{\mathbf{w}}_j$
1: **initialize**: Select $\hat{\mathbf{w}}_j$ in B
2: **for** $t = 1, 2, ..., T$ **do**
3:     $\hat{\mathbf{w}}_j = Avg(w_j + K(w_i))$
4: **end for**

---

4 Experiments

4.1 Experiment Setup

Datasets and Models. We experiment on MNIST, CIFAR10 and TinyImageNet datasets. For MNIST, we use Lenet as backbone. For CIFAR10 and TinyImageNet, we use Resnet18. In addition, we use U-Net to train attack models.

Attack Method. We simulated an attack method called IBA.

Defense Baseline. We choose FedAvg as defense baseline and compare F2L+SOTA with 4 SOTA defenses Krum, NDC, RFA and FLAME.

Evaluation Metrics. We consider a set of metrics for evaluating the effectiveness of backdoor attack and defense techniques as follows:

Main task Accuracy (MA) measures the model's ability to correctly handle input data under normal conditions. The attacker attempts to minimize any changes to this metric, while the defense mechanism must ensure that the accuracy is not compromised.

Backdoor Accuracy (BA) measures the model's ability to produce incorrect outputs when faced with a backdoor attack. The attacker aims to increase this rate, while defense mechanisms strive to reduce it.

4.2 Experiment Result

F2L, as a component integrated with existing defense methods, enhances protection against more covert backdoor attacks.

In Table 1, the significant decrease in BA with F2L+NDC and F2L+RFA is due to the fact that NDC and RFA do not address parameters updated less frequently during global aggregation rounds. This oversight allows covert backdoor attacks like IBA to bypass these defenses. F2L addresses this gap by focusing on fine-tuning these less frequently updated parameters. As a result, when combined with NDC or RFA, F2L significantly reduces the success rate of backdoor attacks.

F2L+FLAME. FLAME consists of three steps: clustering and filtering with HDBSCAN and cosine similarity to remove obvious malicious updates, clipping parameters with L2 norms above a threshold S, and adding Gaussian noise. Integrating F2L between the clustering and clipping steps enhances FLAME's ability to address covert backdoor attacks by fine-tuning less frequently updated parameters, improving overall defense. Integrating F2L between the clustering and clipping steps enhances FLAME's ability to address covert backdoor attacks by fine-tuning less frequently updated parameters, improving overall defense.

From Table 1, it is evident that F2L+SOTA shows a significant improvement in BA compared to SOTA. For example, with NDC, the BA reaches 98.11%, indicating that it cannot effectively defend against covert backdoor attacks like IBA. However, with the introduction of F2L, the BA drops to only 21.32%. Therefore, we can confirm that F2L effectively mitigates the impact of covert backdoor attacks.

Table 1. MA and BA in three datasets.

Method	MNIST		CIFAR10		TinyImageNet	
	MA(%)	BA(%)	MA(%)	BA(%)	MA(%)	BA(%)
FedAvg	98.11	99.07	84.49	87.13	64.36	94.19
Krum	74.22	99.11	56.74	98.22	48.91	97.95
NDC	86.14	98.11	80.22	97.44	63.23	93.46
RFA	69.34	98.79	60.21	98.26	50.87	94.73
FLAME	99.14	19.13	79.35	4.04	66.81	0.56
F2L+NDC	85.84	**21.32**	80.22	**20.16**	63.23	**13.41**
F2L+RFA	68.94	**19.74**	60.21	**9.36**	49.87	**12.53**
F2L+FLAME	99.14	**9.38**	79.35	**0.00**	66.81	**0.00**

We plot data from federated aggregation rounds 500–700 in Fig. 3 and Fig. 4 (dataset: MNIST). Figure 3 compares the MA of F2L combined with NDC, RFA, and FLAME. The solid line represents our work, which maintains MA largely unaffected. This is because gradient projection in F2L minimizes parameter changes from the previous round, keeping MA stable.

Figure 4 shows the BA of F2L+NDC, F2L+RFA, and F2L+FLAME compared to NDC, RFA, and FLAME. F2L's parameter range BBB significantly reduces BA when combined with these defenses. Notably, F2L+NDC and F2L+RFA show the most significant decreases, as NDC and RFA initially did not address less frequently updated parameters during global aggregation.

Runtime Overhead of F2L. We validate F2L's lightweight nature through comparative experiments. We compare the time overhead when deploying F2L

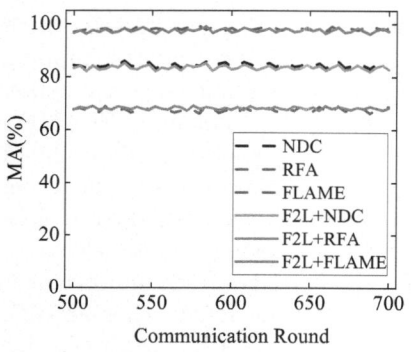

Fig. 3. F2L+SOTA compare with SOTA in MA.

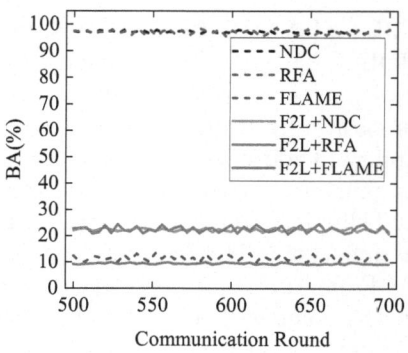

Fig. 4. F2L+SOTA compare with SOTA in BA.

(With F2L) and without deploying F2L (Without F2L). Additionally, we compare F2L's time overhead with other SOTA defense methods to assess its efficiency as a lightweight defense mechanism.

The time overhead of F2L is primarily consumed by the Krum algorithm, which has a complexity of $O(N^2)$, where N is the number of parameters. In comparison, the time overhead for Select w_j and Gradient Projection is only 0.1, which can be considered negligible. The reason is that clients record historical parameter information during training. Gradient masking is used to reduce communication volume. This helps decrease time overhead. For Gradient Projection, the computation is linear with a complexity of $O(N)$, resulting in very low time overhead.

Table 2. Time overhead of components in F2L. G. Projection denotes Gradient Projection.

F2L	Time Overhead (s)
Select w_j	0.1
Krum	5.2
G. Projection	0.1

Table 3. Time overhead of F2L and existing defense SOTA.

Method	Time Overhead (s)
Krum	7.3
NDC	6.7
RFA	5.9
FLAME	6.2
F2L	**5.2**

As shown in Table 3, F2L has a clear advantage in terms of time overhead compared to existing defense methods. First, F2L does not compute the values between each client's uploaded parameters like Krum does. Instead, it only computes the filtered parameters within each client. Therefore, F2L has much lower time overhead compared to Krum. This is also why the Krum in Table 2 has a smaller time overhead compared to the Krum in Table 3. Secondly, compared to NDC, F2L's Select w_j step saves significant time overhead. It is not sensitive to changes in the number of labels. In contrast, NDC's time overhead grows proportionally with the number of labels when handling large-scale label tasks. RFA uses the geometric median method to mitigate backdoor effects. Its operation is simple and flexible, with low time overhead. However, RFA's time overhead also grows proportionally with the number of labels. Finally, FLAME's time overhead is concentrated in the Add Noise step, as the modified parameters need to be retrained to maintain normal performance. Therefore, based on the analysis of Table 3, F2L is indeed a lightweight defense method.

How to Use F2L in Combination with Existing Methods. In this paper, We target the defense against invisible backdoor attacks (e.g., IBA). Unfortunately, existing methods have failed to counter IBA effectively. Therefore, we propose F2L. This section aims to demonstrate the scalability of F2L. Beyond invisible backdoor attacks, we assume a stronger attack scenario that includes both invisible and traditional backdoor designs. For Krum, NDC, and RFA methods, F2L should be deployed beforehand. F2L is a lightweight defense with low time overhead. After F2L processes the parameters, the time overhead for these three methods is significantly reduced (Table 4). This is due to Gradient Projection in F2L, which projects gradient vectors in a specific direction and simplifies the steps for these methods. For the FLAME method, F2L should be placed between Dynamic Clustering and Adaptive Clipping. Dynamic Clustering removes malicious updates with low cosine similarity. In F2L, the Krum step does not need to consider the effect of malicious updates within normal updates. This significantly increases the efficiency of F2L combined with FLAME.

Table 4. F2L combines with 3 methods with different time overheads due to different ordering.

Front—Back	Time Overhead (s)
Krum — F2L	12.5
NDC — F2L	11.9
RFA — F2L	11.1
F2L — Krum	**10.1**
F2L — NDC	**9.2**
F2L — RFA	**10.4**

5   Conclusion

Invisible backdoor attacks in FL are rapidly evolving with high impact, stealthiness, and persistence, and current defenses are inadequate. This paper proposes a Lightweight Focus Layer (F2L) that targets less frequently updated parameters during global aggregation, using Krum and gradient projection for fine-tuning. Combining F2L with existing defenses, our experiments show that our method maintains the original MA while significantly reducing BA. In addition, we validate the lightweight nature of F2L by comparing its advantages in terms of time overhead with existing defense methods.

References

1. McMahan B., Moore E., Ramage D., Hampson S., Aguera y Arcas B.: Communication-efficient learning of deep networks from decentralized data. In: Proceedings of the 20th International Conference on Artificial Intelligence and Statistics, AISTATS 2017, PMLR, Online, vol. 54, pp. 1273–1282
2. Li L., Yuxi F., Kuoyi L.: A Survey on Federated Learning. In: IEEE 16th International Conference on Control & Automation, ICCA 2020, Singapore, pp. 791–796 (2020)
3. Lingjuan L., Han Y., Qiang Y.: Threats to Federated Learning: A Survey. ArXiv, vol. abs/2003.02133 (2020)
4. Priyanka Mary M.: Federated Learning: Opportunities and Challenges. ArXiv, vol. abs/2101.05428 (2021)
5. Bagdasaryan, E.; Veit, A.; Hua, Y.; Estrin, D.; Shmatikov, V.: How to backdoor federated learning. In: Proceedings of the 2020 International Conference on Learning Representations, ICLR 2020, Addis Ababa, Ethiopia, PMLR, Online, vol. 108, pp. 2938–2948
6. Thuy Dung N., Tuan A. N., Anh T., Khoa D.D., Kok-Seng W.: IBA: towards irreversible backdoor attacks in federated learning. In: Proceedings of the 37th International Conference on Neural Information Processing Systems, NeurIPS 2023, Curran Associates, Inc, 2023, vol. 36, pp. 66364–66376
7. Blanchard P., El Mhamdi E. M., Guerraoui R., Stainer J.: Proceedings of the 31st International Conference on Neural Information Processing Systems, NeurIPS 2017, Curran Associates, Inc, vol. 11, pp. 118–128 (2017)
8. Ziteng S., Peter K., Ananda Theertha S.: Can You Really Backdoor Federated Learning? ArXiv, vol. abs/1911.07963 (2019)
9. Dwork, C., Roth, A.: The algorithmic foundations of differential privacy. In: Foundations and Trends in Theoretical Computer Science (2014). https://doi.org/10.1561/0400000042
10. Luis M., Kenneth T.C.o., Emil C.L.: Byzantine-Robust Federated Machine Learning through Adaptive Model Averaging. ArXiv, vol. abs/1909.05125 (2019)
11. El Mahdi El M., Rachid ., Sébastien R.: Proceedings of the 35th International Conference on Machine Learning, ICML 2018, PMLR, Online, vol. 80, pp. 3521–3530
12. Wenke, H., Mang, Y., et al.: Federated learning for generalization, robustness, fairness: a survey and benchmark. IEEE Trans. Pattern Anal. Mach. Intell. (2024). https://doi.org/10.1109/TPAMI.2024.3418862
13. Liu, T., Zhang, Y., Feng, Z., Yang, Z., Xu, C., Man, D., Yang, W.: Beyond traditional threats: a persistent backdoor attack on federated learning. In: Proceedings of the AAAI Conference on Artificial Intelligence, AAAI, vol. 38, pp. 21359–21367 (2024)

14. Pillutla, K., Kakade, S.M., Harchaoui, Z.: Robust aggregation for federated learning. IEEE Trans. Signal Process. **70**, 1142–1154 (2022). https://doi.org/10.1109/TSP.2022.3153135
15. Thien Duc N., Phillip R, et al.: FLAME: taming backdoors in federated learning. In: 31st USENIX Security Symposium, USENIX Security 2022, pp. 1415–1432

Intelligent Backpack Based on Wireless Mobile Technology

Hua Wang[1](✉), Pengxiang Wang[2], Chuyan Wang[1], Zhengsheng Xiao[1], and Yuhong Sun[1]

[1] School of Computer Science, Qufu Normal University, Rizhao 276826, China
wanghua@qfnu.edu.cn
[2] Jinan Foreign Language School, Jinan, China

Abstract. In this paper, we design an intelligent backpack based on wireless mobile technology, which includes a backpack ontology, mobile terminal, and terminal app. On the backpack ontology, there are several related modules such as the master control chip, voice recognition module, wireless transmission module, GPS/GPRS module, and acceleration gyroscope module, among others. To enhance security and privacy, we have integrated a fully anonymous authentication scheme based on hash functions and elliptic curves. This scheme ensures the confidentiality, security, and user's privacy. The master control chip communicates with the terminal app through the wireless transmission module. The backpack can realize wireless network services, multimedia services, security functions, and more. The experiment results show that the intelligent backpack can ensure personal and property safety when the backpack user faces danger. Therefore, the intelligent backpack has strong emergency response capabilities, real-time features, and it has a promising application prospect.

Keywords: wearable device · Bluetooth · Wi-Fi · security function · mobile phone app · Fully Anonymous Authentication

1   Introduction

The advent of the intelligent wave has led to the gradual development of science and technology towards intelligence,digitalization,networking,and humanization.The birth of the smartphone has transformed our lifestyle, work methods, and environment [1]. Concurrently, smart hardware has experienced rapid development, making the interaction design of the human-machine interface increasingly significant. Consequently, numerous smart wearable devices integrated with smartphones have emerged [2], and the global wearable device market has heated up rapidly since 2013. Products such as smart glasses, smartwatches, and smart bracelets have appeared in quick succession. Standalone wearable devices have become the mainstream in the wearable industry, and these devices can interact synergistically with the Internet of Things. Wearable devices are recognized as one of the significant directions for the development of intelligent terminals in the IT industry and are expected to become a leading industry in the future [3,4].

F. Zhang et al. (Eds.): AIS&P 2024, LNCS 15399, pp. 44–58, 2025.
https://doi.org/10.1007/978-981-96-1148-5_5

Wearable devices represent an emerging market with immense potential. As more wearable devices enter the market, they are gradually being accepted by the general public [5]. Wearable devices serve as a new means and entry point for users to access the internet [6]. With the gradual application of technologies such as wireless sensor networks, human-computer interaction, cloud computing, big data, and the Internet of Things, the application services and user experiences of wearable devices are continuously enriched [7]. Currently, many wearable device applications are relatively fixed, with smart wearable devices primarily used in fitness and health monitoring, as well as in medical assistance [8,9]. Entertainment and leisure are also areas where wearable devices are applied [10], and these technologies are quite mature. However, there is a lack of research in the fields of network services and security protection. Therefore, this paper designs a wearable device that integrates various functions such as wireless networking, security,positioning, and voice recognition-the intelligent security backpack.

2   System Overview

2.1   Basic Composition of the System

This intelligent security backpack utilizes wireless mobile technology and comprises a backpack body, mobile terminal,and terminal application within the mobile terminal,which is composed of two components: the user app and the guardian app.The backpack incorporates a main module, a button module, an LED flash module, and an audio module .The main module is composed of a master control chip,a voice recognition sub-module,a wireless transmission sub-module, an accelerometer gyroscope sub-module,a GPS/GPRS sub-module, and an LED lighting indicator sub-module which is made up of the first MOSFET, LED lights and LED driving chips.as well as a power management sub-module.Additionally,the wireless transmission sub-module consists of a Bluetooth secondary sub-module (with audio and data transmission functions) and a WiFi secondary sub-module;The audio module consists of a motor vibration sub-module and a speaker sub-module, where the speaker sub-module consists of a speaker and an audio amplifier circuit;The LED flash module consists of a second MOSFET, an LED flash, and an LED driving chip;The button module has several functional buttons, including a volume control button, an emergency call button,a confirmation button, and a cancellation button.

2.2   Connection Relationships of System Modules

The connection relationships among the system modules are shown in Fig. 1. The audio module is connected to the input/output (IO) port of the master control chip through the motor vibration sub-module.For the LED flash module, one end is connected to the output of the driving chip, and the other end is connected to the master control chip through the second MOSFET. Within the main module, the voice recognition sub-module is connected to the master control chip via the Serial Peripheral Interface(SPI) communication protocol;The accelerometer gyroscope module communicates with the master control chip via the Inter-Integrated Circuit (I2C) bus; the GPS/GPRS submodule communicates with the master control chip through the Universal Synchronous/Asynchronous Receiver/Transmitter (USART). In the LED lighting indicator sub-module, one terminal of the LED light is connected to the output of the driving chip, while the other terminal is connected to the master control chip via the first MOS-FET. Drain (D) is connected to one end of the LED light, and the gate (G) is connected to an input/output (IO) port of the master control chip, enabling the chip to regulate the brightness of the light through Pulse Width Modulation (PWM). The master control chip communicates with the Bluetooth secondary sub-module and the Wi-Fi secondary sub-module of the wireless module using the USART. The Bluetooth secondary sub-module establishes wireless connections with mobile devices to transmit data and audio signals, while the Wi-Fi secondary sub-module connects to Wi-Fi enabled devices for data transfer. The power management module supplies power to all modules.

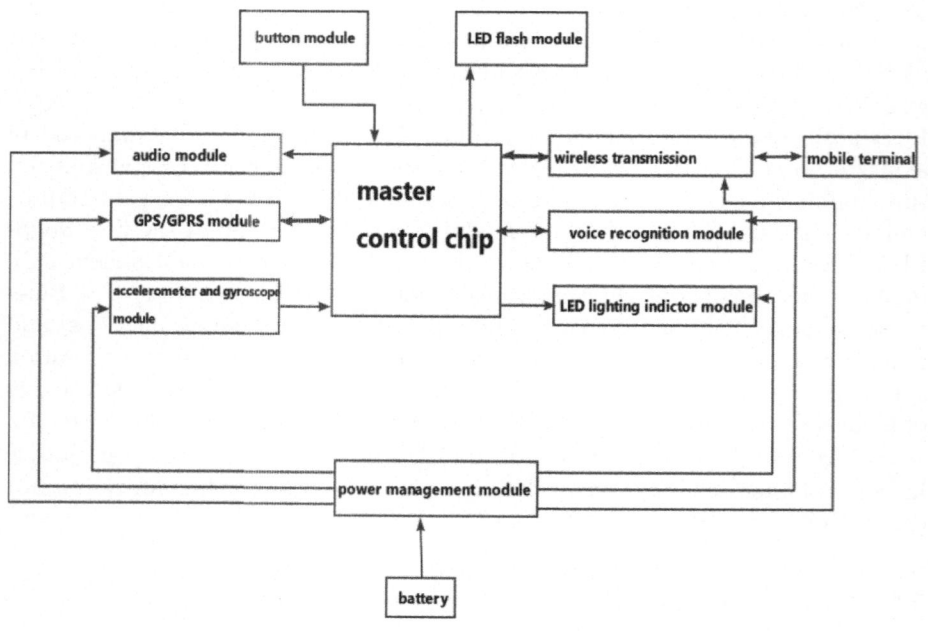

Fig. 1. system module connection diagram.

2.3   System Module Integration Relationship and Spatial Distribution

The distribution of system modules on the backpack is shown in Fig. 2. In this design, the power management sub-module,master control chip,voice recognition sub-module,wireless transmission sub-module, GPS/GPRS sub-module, LED lighting indicator sub-module and accelerometer gyroscope sub-module, as well as peripheral interface circuit are integrated on the main printed circuit board (PCB). The main control PCB is situated at the front of the backpack's shoulder strap.The LED flash module is integrated onto an another PCB, positioned at the rear of the backpack's shoulder strap.The audio module is integrated onto a PCB and is situated in the upper region of the backpack's shoulder strap.The button module is positioned at the terminal end of the elastic band of the backpack. As shown in Fig. 2, the LED flash module, audio module, main module, and button module are packaged with waterproof materials and fixed to their respective positions on the backpack shoulder straps through nylon ties.The modules mentioned previously are linked through the internal wiring integrated into the shoulder straps.

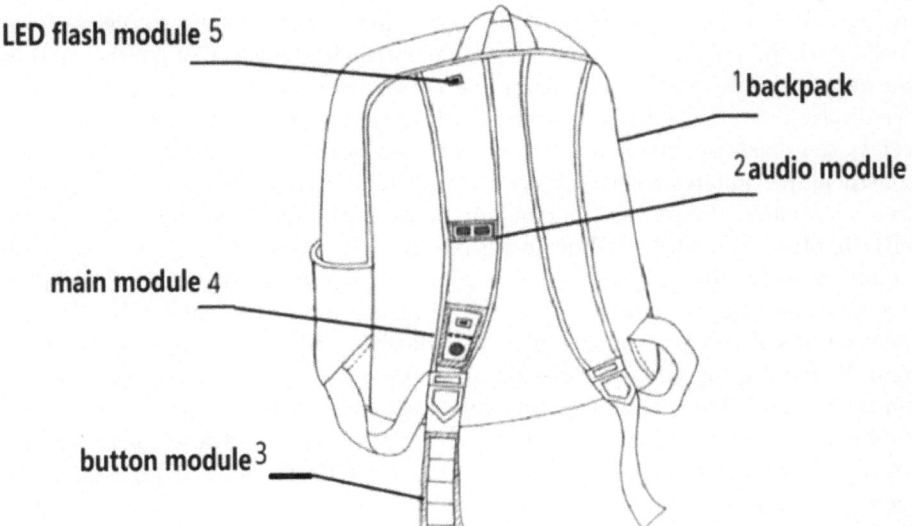

Fig. 2. System Module Layout Diagram

3   System Functionality Design

3.1   System Function Overview

The intelligent security backpack designed in this article has nine fundamental functionalities: voice recognition, security,communicate security function, network services, anti-loss positioning, motion state monitoring,hands-free phone calls, Bluetooth speaker,and lighting.

Voice Recognition Function: The voice recognition module processes the acquired voice signals and compares them against those in the speech database. Upon a successful match, the corresponding serial number is transmitted to the master control chip via SPI communication protocol.The master control chip then evaluates the received serial number and executes the corresponding operation based on the serial number.This function primarily serves to regulate the activation and deactivation of other functions, and is meaningful only when combined with other functionalities.

Security Function: The flowchart of the security function execution is shown in Fig. 3.When the users are in danger, they may either vocalize a distress call, such as 'Help,' or press the button module on the backpack. Upon either action, the backpack's security function is triggered.The voice recognition module identifies distress vocalizations and subsequently transmits the recognized serial number to the master control chip. Upon reception of the serial number, the master control chip evaluates whether it corresponds to a distress command. If it is a distress command, the master control chip controls the LED flash module and the LED lighting indicator module to initiate a continuous flashing pattern, while simultaneously directing the speaker module to produce an alarm sound.The strong flashing effect can temporarily blind the criminals, thus achieving the purpose of self rescue. Concurrently, the master control chip activates the GPS module, facilitating the transmission of location data to the master control chip via the USART.The master control chip dispatches the information to the GPRS module via the USART, which subsequently forwards the data to the server utilizing the Transmission Control Protocol/Internet Protocol (TCP/IP) suite.The server transmits the location data to the respective guardian's mobile application via Socket connections, in accordance with the established binding relationships between the user and the guardian.Furthermore, the main control chip relays an emergency command to the Bluetooth sub-module via the USART. Subsequently, the Bluetooth sub-module sends the command to the user's mobile device through the wireless connection that has been established with the phone's Bluetooth functionality. The user mobile app transmits text messages to the configured guardianship group, The guardian app of the guardian terminal receives and detects the message content dispatched by the user's app.If the message content corresponds to a pre-defined distress signal,the system autonomously initiates a sequence of actions on the guardian's terminal: it triggers continuous vibration and audible alerts, and seamlessly accesses the map location display interface, thereby facilitating the guardian's ability to promptly ascertain the rescuer's location .

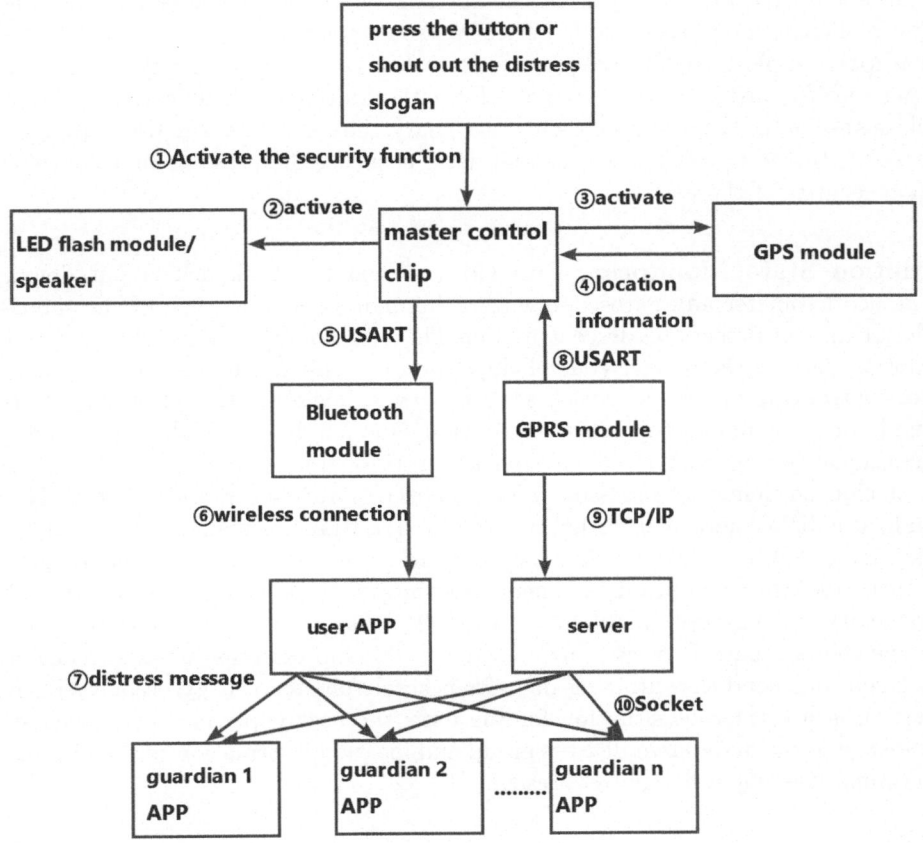

Fig. 3. Security Feature Process Flowchart

Network Service Function: This backpack utilizes WiFi and Peer-to-Peer (P2P) technologies to facilitate close-range social interactions, enable file transfers,and shares WiFi network hotspots through 3G/4G mobile networks. In scenarios where external network access is not available, the backpack can scan for other users via WiFi and retrieve public information(such as age, gender, and hobbies), which is then uploaded to the mobile app via Bluetooth.Utilizing their mobile devices, users can access the public information of nearby individuals. Furthermore, by using WiFi technology, they can establish P2P connections with other users, thereby facilitating conversations and wirelessly transfer files within close range.

Communicate Security Function: The intelligent backpack integrates a fully anonymous identity authentication scheme between the backpack and the mobile terminal app. By employing pseudonym-based private keys and actual private keys, the authentication process between the backpack user and the backpack can be rendered fully anonymous, thereby resisting various attacks and preventing malicious interference and theft of personal information. Additionally, the backpack employs conditional privacy protection, allowing the backpack and the user to safeguard sensitive personal information during communication and disclose specific information only when necessary. This balances the user's privacy needs with the utilization of essential data, ensuring the personal and information security of the user.

Motion State Monitoring Function: When the backpack is put down, the accelerometer and gyroscope within the main control module of the shoulder strap can detect the descent motion.These sensors then relay the acquired motion data to the master control chip.Upon recognizing the motion, the master control chip orchestrates the backpack to autonomously transition into sleep mode.In sleep mode, the voice recognition functionality and Bluetooth audio functionality are deactivated, the speaker stops operation, and the master control chip suspends its assessment of potential hazardous motion states. The indicator lights and lighting lights integrated within the main module will be deactivated.When the user lifts the backpack, the corresponding functionality is activated,thereby optimizing energy consumption and enhancing operational longevity.Furthermore, in the event of a fall, collision, or inadvertent dropping while the backpack is in use, the master control chip discerns the emergence of a hazardous condition utilizing data from accelerometer and gyroscope .When detecting a dangerous situation for the user, the user app will send a distress message to the designated distress group and location information to the distress terminal, thereby achieving self-rescue.

Bluetooth Audio Function: There is a stereo speaker on the upper part of the shoulder strap of the backpack. Once the user's phone is connected to the backpack's audio Bluetooth, the audio from the phone will be played through the backpack's speaker via Bluetooth. The user can control the playback status of the phone music through voice control ,while the volume can be precisely adjusted through button module.

Hands-Free Calls Function: Upon receiving an incoming call, the phone will facilitate the transmission of call-related data to the master control chip via Bluetooth.The master control chip controls the speaker and motor vibration module to generate audio and vibration alerts, reminding users to answer the call. Users can accept or reject calls through voice control.Once the call is connected,users can adjust the volume via the button module. Furthermore, the backpack features an advanced command control that enables the mobile phone to automatically dial a specified contact or redial the last number, with the entire process controlled by voice commands, completely liberating the user's hands.

Anti-loss Positioning Function: When the mobile app initiates this function,a wireless communication link is established between the user and the backpack's wireless transmission module via the terminal's integrated wireless interface.The integrated Analog-to-Digital (AD) conversion module facilitates the digitization of the signal, then transmits the digital signal to the master control chip via the internal bus; Upon the detection of diminished signal strength by the master control chip, it activates the sound module to generate both auditory and vibration. Simultaneously,the user terminal app orchestrates the mobile terminal to emit alert sound and initiate vibration, bi-directional reminding the user that their phone and backpack have been stolen or forgotten.Once connected to the user's smartphone, the user can track the backpack's location via device's app.In the event of theft, the user can locate the backpack location information through the mobile phone that has been paired with it, and then track the backpack location to assist the police in solving the case. For users who are children or the elderly,family members can bind their mobile phones and backpacks together to monitor the real-time location data of the child or elderly individual through their smartphones.

Night Lighting Function: The backpack's main control module is equipped with LED lights for illumination.Users can control the LED light's on/off state and brightness level through voice commands or the mobile app, achieving the purpose of lighting in the dark.

4 Terminal Application Function Design

4.1 User App

Users can configure the activation and deactivation of the security function , positioning anti-loss prevention function, motion state monitoring function, and other functions through the user app. Upon the user's deactivation of a specific function,the app transmits a corresponding command to the backpack's Bluetooth sub-module via the Bluetooth to turn off the function.The master control chip, after receiving the command to disable the function via the USART, will block the activation path and will no longer respond to such function .

Users can establish a designated group for the transmission of emergency notifications and to specify a contact individual for initiating distress calls via the application interface in the event of an emergency.

In the event of theft of the user's backpack,the user can check the location of the backpack through the user's app. The specific process for querying the back-pack's location is shown in Fig. 4. The user's app sends a request to the server via a socket connection to retrieve the backpack's location data, which is then relayed by the server to the designated backpack's GPRS module utilizing the TCP/IP protocol.The backpack's master control chip receives a location query request via the USART,initiating the GPS positioning module.Thereafter, the GPS module sends the location data back to the master control chip, also utiliz-ing the USART.The master control chip of the backpack transmits the location data to the server via the GPRS module, which then facilitates the forward-ing of this information to the smartphone that is paired with the backpack.The user app retrieves the map display interface based on location information and display the location of the backpack.

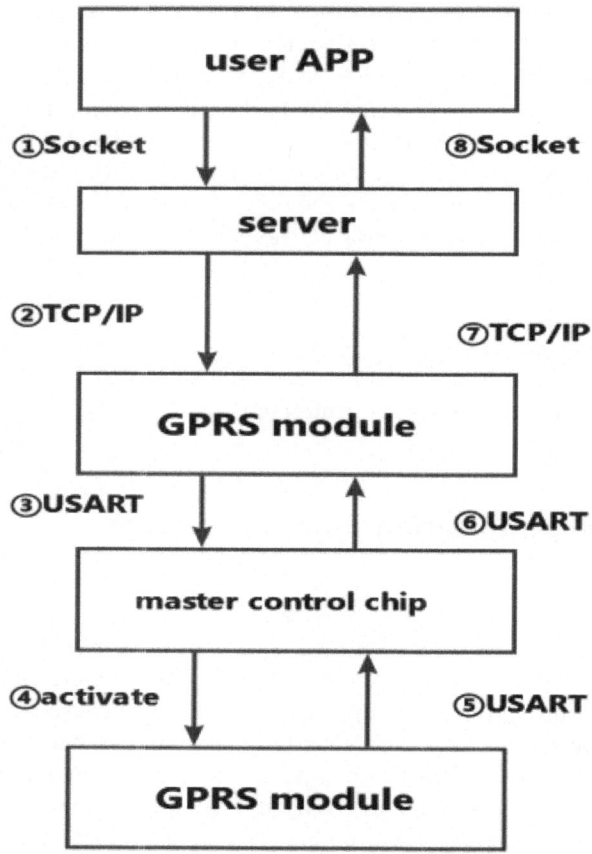

Fig. 4. Illustration of the knapsack Location Query Process

4.2   Guardian App

Upon receiving the consent, the guardian app can view the user's location details through its interface.In the event of an emergency,the user app autonomously initiates an alert message to the designated guardian.Upon receipt of the distress message,the guardian app automatically controls the phone to continuously vibrate and sound reminder alarms,until the guardian confirms the distress message.Subsequently,it automatically retrieves the location display interface, displaying the location details of the rescuer's backpack.

4.3   Fully Anonymous Authentication Scheme

Additionally, we propose a fully anonymous identity authentication scheme between the backpack and the user's end in the communication system. This scheme primarily involves three types of entities: the backpack, the user's mobile app, and the Trusted Third Party (TA). Users and the backpack register with the TA to obtain pseudonym-based private keys, which they then use to compute their actual private keys. Mutual identity authentication is performed by both parties based on their respective public and private keys. Upon successful authentication, they obtain a session key that enables secure communication between the user's app and the backpack.

The fully anonymous authentication requires three processes, and the detailed authentication procedure is as follows:

A. System Initialization Phase: TA selects a random number $s \in Z_q^*$ as the primary key and stores it secretly. Let G_1 and G_2 be two additive cyclic groups of prime order q, and P is the generator of G_1. TA computes the system public key $P_{\mathrm{pub}} = sP$ and selects three hash functions: $H_1 : \{0,1\}^* \rightarrow G_1 \times G_1$, $H_2 : G_1 \times G_1 \rightarrow Z_q^*$, $H_3 : G_1 \times \{0,1\}^* \rightarrow G_1 \times Z_q^*$. e is the bilinear mapping of $G_1 \times G_1 \rightarrow G_2$. The TA publishes the system parameters as param = $\{G_1, G_2, q, P, H_1, H_2, H_3, e, P_{\mathrm{pub}}\}$.

B. Anonymous Registration Phase: During the anonymous registration phase, the users and the backpack apply for registration with TA in the form of pseudonyms. TA generates pseudonym-based private keys for them to complete the registration.

(1) User Registration: The user selects random numbers $r_c, x_c, z_c \in Z_q^*$ and calculates $R_C = r_c P$, $Z_C = z_c \cdot ID_c$, the public key $PK_c = H_1(ID_c, R_C)$, and pseudonym $PID_c = x_c \cdot PK_c$. The user sends the R_C, Z_C, and PID_c to TA and does not need to wait for a reply from TA. The user calculates the session key with TA as $K_{CA} = H_2(PID_c \cdot G_c)$, where $G_c = e(r_c \cdot PID_c, P_{\mathrm{pub}})$.After receiving the message from the user, TA keeps Z_C to trace the real identity of the user. Then, the TA generates a private key $PSK_C = s \cdot PID_c$ based on the pseudonym and calculates the session key $K_{AC} = H_2(PID_c \cdot G_c)$,

where $G_c = e(PSK_C, R_C)$. The TA encrypts the pseudonym-based private key with the session key and sends $I = E_{K_{AC}}(PSK_C)$ to the user. The user decrypts the message using the session key and obtains the pseudonym-based private key as $DK_{AC}(I) = PSK_C$. The user then calculates the real identity-based private key as $SK_C = x_c^{-1} \cdot PSK_C = s \cdot PK_c$. After successful registration, the user obtains the pseudonym-based private key and further calculates the real identity-based private key, which is known only to the user and is used for subsequent identity authentication with the backpack associated with the backpack.

(2) Backpack Registration: The backpack selects random numbers $r_S, x_S \in Z_q^*$ and calculates $R_S = r_S P$, the public key $PK_S = H_1(IDS, R_S)$, and pseudonym $PID_S = x_S \cdot PK_S$. The backpack sends R_S and PID_S to TA and does not need to wait for a reply from TA. The backpack calculates the session key with TA as $K_{SA} = H_2(PID_S \cdot G_S)$, where $G_S = e(r_S \cdot PID_S, P_{\text{pub}})$. After receiving the message from the backpack, TA generates a private key $PSK_S = s \cdot PID_S$ based on the pseudonym and calculates the session key $K_{AS} = H_2(PID_S \cdot G_S)$, where $G_S = e(PSK_S, R_S)$. The TA encrypts the pseudonym-based private key with the session key and sends $I = E_{K_{AS}}(PSK_S)$ to the backpack. After receiving the message, the backpack decrypts it using the session key to obtain $DK_{AS}(I) = PSK_S$. The backpack then calculates the real identity-based private key $SK_S = x_S^{-1} \cdot PSK_S = s \cdot PK_S$. After successful registration, the backpack obtains the pseudonym-based private key and further calculates the real identity-based private key, which is known only to the backpack and is used for subsequent identity authentication with the user.

C. Identity Authentication Phase: In this phase, the user authenticates with the backpack. The user sends the signed information to the backpack for identity verification. Similarly, the backpack sends the signed information to the user for verification. After successful verification, a session key is obtained for subsequent communication.

The user selects a random number $a \in Z_q^*$, calculates $MC = aPK_c$, $NC = aP$, selects a timestamp T_1, and calculates $h_1 = H_3(MC, T_1, NC)$. The signature is computed as $SC = SK_c + (a + h_1)MC$. The user sends the parameters zc (The random number chosen by the user during the registration stage), MC, NC, SC, h_1, T_1, and its public key PK_c to the backpack S.

After receiving the message from the user, the backpack checks the validity of the timestamp. If the timestamp is invalid, the request is rejected. If the timestamp is valid, the backpack verifies the validity of the signature by checking whether $e(P, SC) = e(P_{\text{pub}}, PK_c) \cdot e(NC, MC) \cdot e(h_1 P, MC)$ is equal. If it is equal, the identity authentication is successful, and the backpack calculates the session key and keeps the random number zc for tracing users. Otherwise, the authentication fails.

The backpack selects a random number $b \in Z_q^*$, calculates $MSC = bPK_S$, $NSC = bP$, selects a timestamp T_2, and calculates $h_2 = H_3(MSC, T_2, NSC)$.

The signature is computed as $SSC = SK_S + (b + h_2)MSC$. The backpack sends the parameters MSC, NSC, SSC, h_2, timestamp T_2, and its public key PK_S to the user. At the same time, the backpack calculates the session key as $SSK_{CS} = bNSC + abP$.

After receiving the message from the backpack, the user first checks the validity of the timestamp. If the timestamp is invalid, the request is rejected. If the timestamp is valid, the user verifies the validity of the signature by checking whether $e(P, SC) = e(P_{\mathrm{pub}}, PK_S) \cdot e(NSC, MSC) \cdot e(h_2 P, MSC)$ is equal. If it is equal, the identity authentication is successful, and the user calculates the session key as $SSK_{CS} = aNS + abP$.

At this point, the user and the backpack have authenticated each other's identities, and they can use the session key to encrypt information for subsequent data transmission. The identity authentication process between the user and the backpack is shown in the figure.

Correctness Analysis: During the registration phase, the backpack and TA ensure the correctness of the session key by independently calculating G_c. In the identity authentication phase, both parties verify each other's identity through signature validation; if the signature is valid, they continue communication, otherwise, they terminate the connection, thereby proving their ability to successfully negotiate a session key. Additionally, this paper introduces conditional privacy protection, where the backpack encrypts personal information with a random number during the registration process, which can only be decrypted under specific circumstances, ensuring the privacy and security of the backpack.

Efficiency Analysis: We also discussed the computational overhead of several identity authentication and key agreement protocols. The schemes proposed by Qi and others, as well as by Isalm and others, have higher computational costs due to the use of elliptic curve point multiplication operations. Tsai and others' scheme uses hash and XOR encryption, thus having lower computational costs. Our scheme has the lowest computational cost and can resist various attacks, offering better anonymity (Table 1).

Table 1. Efficiency Analysis Comparison Table

	Isalm et al. [11]'s Scheme	Qi et al. [12]'s Scheme	Tsai et al. [13]'s Scheme	Our Scheme
Registration Phase	5PM+3PA+2H	4PM+1PA+2H	3PM+2PA+1H	1PM+3H
Mutual Authentication Phase	8PM+10H	7PM+1PA+10H	6PM+4PA+8H	6PM+2PA+2H
Total Cost	13PM+3PA+12H	11PM+2PA+12H	9PM+6PA+9H	7PM+2PA+5H

5   Hardware Design and Implementation

In terms of power supply management, the backpack finally uses a 3.7V lithium battery for power supply, and designs for 3.3V and 5.0V DC power. The 3.3V power supply uses the ASM1117-3.3 linear regulator, while the 5.0V power supply uses the FP6291 boost IC, both of which have protection functions. The power supply circuits also include bypass capacitors and decoupling capacitors to improve power stability.

Regarding the main control module, the text compares the STC51 microcontroller with the STM32 microcontroller and ultimately selects the STM32F103ZET6 microcontroller as the main control chip. The STM32F103 has a wealth of on-chip resources and a high operating frequency, making it suitable for the needs of the intelligent backpack system. The role of the STM32F103 in the system includes:

1. Communicating with the voice recognition module via SPI to recognize voice commands. 2. Communicating with the GPS/GPRS module via USART to control the module to return or send positioning data. 3. Communicating with the Bluetooth module via USART to achieve data transmission with mobile phones. 4. Communicating with the gyroscope and accelerometer module via I2C to obtain and process motion information. 5. Reading button status via IO ports and making corresponding responses. 6. Controlling the brightness of the LED module via IO ports.

6   Analysis of Technical Advantages

6.1   Security Function

The backpack features dual activation for its security functions through voice command and button press, enhancing the responsiveness.In emergencies, users can rapidly activate the front and rear shoulder strap flashlights via voice command to produce a strong flash effect for self-protection.It is more timely than traditional self-defense tools,while some self-defense tools may not be allowed through transportation security checks, the flash light of this technology will not face such issues.Additionally, when a dangerous situation occurs, the backpack automatically controls the phone to send out a group of distress messages, thereby offering greater security assurance.

6.2   Communication Security Function

In the communication process between the user and the backpack, in order to fully protect user privacy and maximize the functionality of the backpack, we have proposed a fully anonymous authentication scheme. By introducing a third-party trusted institution, we have sequentially passed through stages such as information registration, identity authentication, and communication, achieving mutual authentication and data sharing between the backpack and the user.

6.3   Voice Recognition Function

The voice recognition commands can be tailored through the user app,making it more humanized.The voice command control for hands-free operation and dialing designated contacts completely frees up the hands, also enhancing the safety of using a mobile phone while the user is on the move.

6.4 Module Placement

This smart backpack features innovative module design and arrangement, with a focus on enhancing user convenience.When the user puts on the backpack,the flashlight installed on the back of the shoulder strap,that is, on the user's back, is specially used to prevent attacks from thugs; the speaker installed above the shoulder strap, that is, directly above the user's shoulder, is close to the user'sears and is used for playing music and answering phone calls; the button module installed at the end of the shoulder strap's elastic band makes it convenient for the user to control the volume and activate security functions, to confirm or cancel certain operations of the user; the main control module installed in front of the shoulder strap, that is, on the user's chest, is the core of the entire system, integrating the acceleration gyroscope, Bluetooth, flash light, main control chip, GPS /GPRS, voice recognition, power management, and vibration motor module on a single PCB board. The flash light module is used to prevent attacks from thugs in front and can also be used as an illumination lighting with adjustable brightness through the mobile app. The various modules are centrally placed on a shoulder strap, which can reduce the length of the internal wiring of the backpack, and each module is detachable, making it easy for later maintenance and cleaning of the backpack. In addition, users can freely adjust the position of the modules with nylon straps,increasing the user's humanized experience.

7 Conclusion

This article designs an intelligent backpack based on wireless mobile technology, which incorporates nine basic functions: voice recognition, security, networking services,communication security function, anti-loss positioning, fitness monitoring, hands-free calling, Bluetooth speaker functionality, and lighting. In the event of a dangerous situation, additional features such as emergency flashlights, mass emergency messaging, and rapid location tracking of the user in distress provide enhanced safety measures. The activation of various functions through voice control offers convenience and immediacy, and the placement of each module has been carefully considered for ease of use by the wearer. This backpack leverages the advantages of wireless mobile technology to realize applications of wearable devices in personal safety, property security, network services, and mobile phone assistance. In addition, the fully anonymous authentication during the communication process between the user and the backpack fully protects the user's privacy, and it surpasses the majority of protective mechanisms in terms of performance and power consumption.It has a broad application audience and can be tailored to meet the specific needs of different user groups, designing personalized functions to satisfy the unique requirements of various populations.

References

1. Raza, A., Al Nasar, M.R., Hanandeh, E.S., Zitar, R.A., Nasereddin, A.Y., Abualigah, L.: A novel methodology for human kinematics motion detection based on smartphones sensor data using artificial intelligence. Technologies **11**(2) (2023)
2. Musci, M., De Martini, D., Blago, N., Facchinetti, T., Piastra, M.: Online fall detection using recurrent neural networks on smart wearable devices. IEEE Trans. Emerg. Top. Comput. **9**(3), 1276–1289 (2020)
3. Park, E.: User acceptance of smart wearable devices: an expectation-confirmation model approach. Telematics Inform. **47**, 101318 (2020)
4. Zovko, K., Šerić, L., Perković, T., Belani, H., Šolić, P.: IoT and health monitoring wearable devices as enabling technologies for sustainable enhancement of life quality in smart environments. J. Clean. Prod. **413**, 137506 (2023)
5. Poongodi, T., Krishnamurthi, R., Indrakumari, R., Suresh, P., Balusamy, B.: Wearable devices and IoT. In: A Handbook of Internet of Things in Biomedical and Cyber Physical System, pp. 245–273 (2020)
6. Nahavandi, D., Alizadehsani, R., Khosravi, A., Acharya, U.R.: Application of artificial intelligence in wearable devices: opportunities and challenges. Comput. Methods Programs Biomed. **213**, 106541 (2022)
7. Gopalakrishnan, M.A., et al.: AI based smart wearable safety system for women to fight against sexual assault and harassment with IoT connectivity. In: AIP Conference Proceedings, vol. 2790. AIP Publishing (2023)
8. Al Hassani, R.T., Atilla, D.C.: Human activity detection using smart wearable sensing devices with feed forward neural networks and PSO. Appl. Sci. **13**(6), 3716 (2023)
9. Nguyen, T., Tran, K.D., Raza, A., Nguyen, Q.T., Bui, H.M., Tran, K.P.: Wearable technology for smart manufacturing in industry 5.0. In: Artificial Intelligence for Smart Manufacturing: Methods, Applications, and Challenges, pp. 225–254. Springer (2023)
10. Bellavista-Parent, V., Torres-Sospedra, J., Perez-Navarro, A.: New trends in indoor positioning based on WiFi and machine learning: a systematic review. In: 2021 International Conference on Indoor Positioning and Indoor Navigation (IPIN), pp. 1–8. IEEE (2021)
11. Islam, S., Biswas, G.: An improved ID-based client authentication with key agreement scheme on ECC for mobile client-server environments. Theor. Appl. Inf. **24**(04), 293–312 (2012)
12. Qi, Y., Tang, C., Xu, M., et al.: An identity-based mutual authentication with key agreement scheme for mobile client-server environment. In: Communications Security Conference, pp. 1–5. IEEE (2014)
13. Lu, Y., Li, L., Peng, H., et al.: Robust anonymous two-factor authenticated key exchange scheme for mobile client-server environment. Secur. Commun. Netw. **9**(11), 1331–1339 (2016)

Industry Upgrading and Public Opinion Prevention Using BERTopic

Xinhua Wang[1], Xiaomei Yu[1,2], Yunmeng Jiang[1], Xiangwei Zheng[1,2], and Wei Li[3(✉)]

[1] School of Information Science and Engineering, Shandong Normal University, Jinan, China
[2] Shandong Provincial Key Laboratory for Distributed Computer Software Novel Technology, Jinan 250358, China
[3] Shandong Normal University Library, Shandong Normal University, Jinan, China
liww72@sdnu.edu.cn

Abstract. The development status of the hotel industry in a city is also an essential factor influencing local tourism revenue. In the tourism industry, besides dining and attractions, which are hotspots for public opinion, the hotel industry is also a key area. The development status of the hotel industry in a city is also an essential factor influencing local tourism revenue. In this context, it is crucial for hotels to address how to prevent such negative public opinion crises. In this paper, we use hotel review texts as an example to discuss how to use BERTopic and GRU networks to extract topics and analyze sentiment from guest reviews. This analysis helps understand guests' focus and sentiment trends regarding hotels, enabling better grasp public opinion direction and user needs. Through topic extraction and sentiment analysis, we can extract crucial information from large scale hotel review texts. This enables monitor public opinion to help prevent large scale public opinion crises, allows for targeted improvements in the hotel industry, and supports decision-making for hotels and the tourism industry.

Keywords: BERTopic · Gated Recurrent Unit · Sentiment analysis · Public opinion management · topic clustering

1 Introduction

Compared to other channels, hotels pay more frequent attention to reviews on Online Travel Agencies (OTAs). Most hotels set a maximum response time for OTA reviews as a performance evaluation metric for staff, so feedback on OTAs is relatively prompt. However, hotels tend to show less attention to reviews on social media platforms like Weibo and Xiaohongshu. For a resort hotel, the driving effect of these channels is particularly significant. This is because when choosing a hotel for business, guests often focus on budget and location. However, when selecting a resort hotel, guests typically gather information from preferred websites, apps, or travel-related Key Opinion Leaders (KOLs). If hotels

F. Zhang et al. (Eds.): AIS&P 2024, LNCS 15399, pp. 59–73, 2025.
https://doi.org/10.1007/978-981-96-1148-5_6

overlook negative feedback on these channels, it can impact their reputation. However, manually processing the reviews text from these platforms is difficult. This paper discusses the information extraction capabilities of BERTopic and Gated Recurrent Unit (GRU) for review texts. GRU is a simplified version of Long Short Term Memory (LSTM), reducing the number of parameters, making model training quicker. Using GRU for sentiment analysis is an efficient and mature method. BERTopic, based on the Transformer and pre-trained BERT models, has strong topic extraction capabilities. This paper innovatively achieves the following points:

1. Collected user reviews from websites like Ctrip, Zhixing, and Xiaohongshu using keywords related to hotel and destination names, creating a dataset on tourists' accommodation experiences and performing information mining.
2. Applied the deep learning-based BERTopic model for topics extraction and visualization from the hotel review datasets, providing reliable basis for hotel upgradation and public sentiment monitoring.
3. Utilized GRU to assess user experience and monitor the trend of hotel public sentiment.

2   Related Work

2.1   Research on Travel Review Texts Analysis

Currently, machine learning and deep learning are extensively utilized in text analysis within the tourism industry. For example, Tianyu Jia and Chen Zhang developed a CNN-BiLSTM-ATT model to explore tourists' dual-factor experience features at hotels [1]. Xin Li and Ying Wang, using the example of predicting visitor flow in Jiuzhaigou Valley, employed fine-grained sentiment analysis based on machine learning to analyze text from multiple sources such as Ctrip, Qunar, Dianping, and Meituans [2]. Qian Zhou and Lei Jiang proposed a new method for sentiment analysis to understand and analyze online public sentiment. They first constructed a sentiment lexicon reflecting semantic features of hotel reviews, then expanded the sentiment words, and finally used deep learning to classify the reviews [3]. Experiments analyzing public sentiment from travel hotel reviews in Xiangtan City revealed that this method provides a more accurate representation of consumer sentiment towards travel hotels, uncovering each hotel's unique features and management shortcomings, thereby supporting decision-making.

2.2   Research on Sentiment Analysis

Currently, text sentiment analysis commonly employs three approaches: sentiment dictionary methods, machine learning, and deep learning. Methods based on sentiment dictionary require a comprehensive and accurate sentiment dictionary. Early research on this approach was conducted by scholars abroad, such as Huettner, who developed a sentiment dictionary with categories and intensity, and expanded the sentiment set through a fuzzy lexicon [4]. Common machine learning methods include Maximum Entropy, Naive Bayes, and so on. Pang categorized movie reviews as positive and negative sentiments [5]. Yang Zhao and others used K-means for clustering user reviews from a shopping app, extracted feature words as indicators of user satisfaction, and used a CNN-SVM model to compute sentiment indicators and assess user satisfaction [6]. Deep learning methods do not require human intervention, as they automatically perform feature extraction and selection, making them more versatile compared to the aforementioned methods. Currently, deep learning is dominant. Recurrent Neural Networks (RNNs) are mainly categorized into three types: RNNs; LSTM; GRU. In 2014, Kim and others used Convolutional Neural Networks (CNNs) for classifying English content and created an improved model based on this approach [7].

2.3   Research on Topic Extraction

Topic models are used to extract topics from large of texts. The core idea is that each document can be viewed as a mixture of multiple topics, with each topic consisting of a set of words. Common algorithms include LSA, PLSA, and LDA. Wang Wei, Gao Ning, Xu Yuting, and others have used the LDA topic model to analyze the dynamic evolution of topics in crowd funding project reviews, finding that LDA is effective for handling the evolution of topics in online reviews [8]. BERTopic offers high accuracy and interpretability in topic modeling and performs better with short texts.

3   Methods

3.1   Gated Recurrent Unit

The GRU is an variant of LSTM, introduced by Cho et al. in 2014s [9]. Compared to LSTM networks, GRU has fewer parameters [10], which results in shorter training times's [11]. GRU uses only two gates-the reset gate and the update gate-which help retain long-sequence text information and pass it to the next neural unit without being removed over time, thereby avoiding the vanishing gradient and addressing long-term dependencies. Compared to LSTM, GRU is more efficient, as illustrated in the structure shown in Fig. 1.

1) Update gate: It regulates how much of the previous state ht-1 is retained in the new state ht. A higher value of the update gate means that more information from the previous state is preserved.
2) Reset gate: It forces the hidden state to discard information unrelated to the prediction while utilizing current input to build more relevant features. The core idea is to determine which information needs to be forgotten.

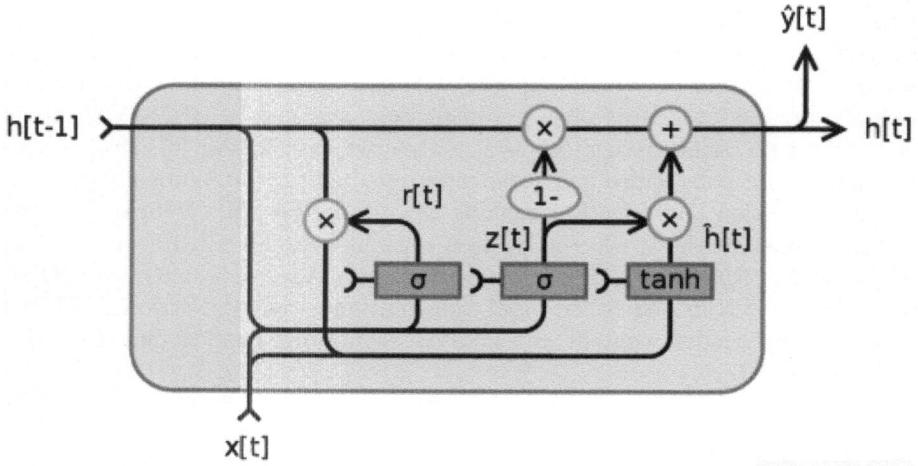

Fig. 1. Schematic representation of the GRU network model.

3.2 Topic Modeling Based on BERTopic

The BERTopic involves 5 steps,which are shown in Fig. 2.
Text Embedding: Utilize the pre-trained language model to convert texts to vectors.
Dimensionality Reduction: Apply techniques like Uniform Manifold Approximation and Projection, Principal Component Analysis and others for dimensionality reduction.
Clustering: BERTopic uses the Hierarchical Density-Based Spatial Clustering of Applications with Noise to cluster document embeddings [12]. In addition to this, K-means algorithm also performs well in clustering.
Weighting Scheme: Use the term frequency-inverse document frequency, the Best Matching 25(BM25) algorithm can also be used.
Topic Selection and Aggregation: BERTopic applies Maximal Marginal Relevance (MMR) to select relevant topics, remove irrelevant ones, and control topic numbers by setting a threshold.

4 Experiment

4.1 Experimental Environment and the Datasets

The experiments were performed under a Windows11 (64-bit) system with a CPU of Intel (R) Core (TM) i5-8265U CPU @ 1.60 GHz, 1.80 GHz, using Python 3.9 and a deep learning framework for Pytorch 1.8. The table below illustrates the experimental environment.

The data set is to crawl the posts and comments of Xiaohongshu and Ctrip through Python crawler technology, and then clean the climbing results and manually mark them to build the required hotel comment data set of tourists.

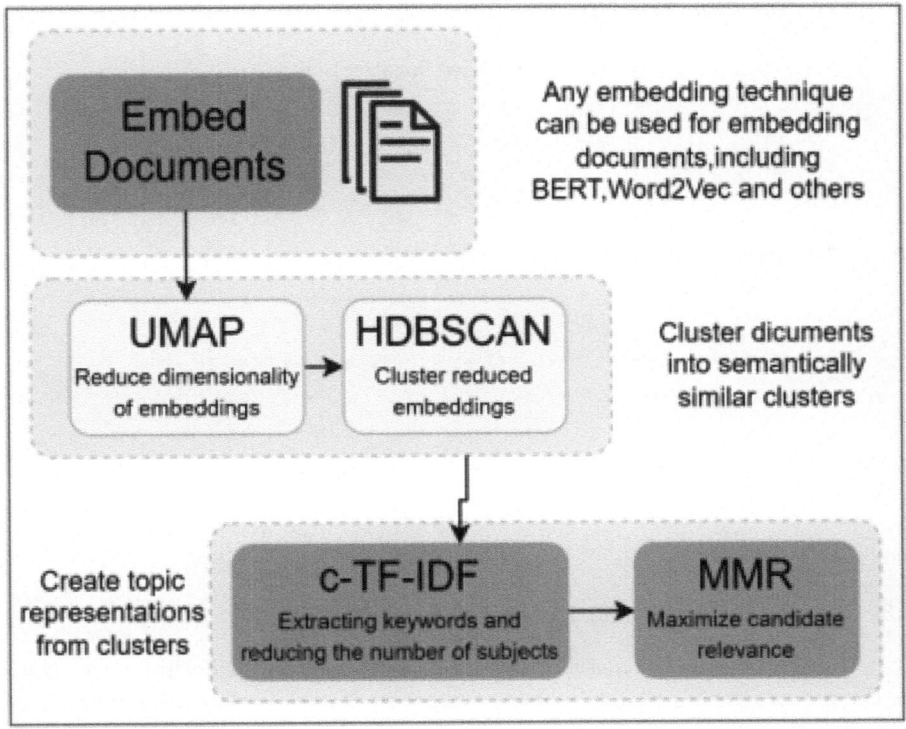

Fig. 2. The algorithm flow chart of BERTopic.

Table 1. Description of the experimental environment.

Experimental environment	Configure
Operating System	Windows11(64bit)
CPU	Intel(R) Core(TM) i5-8265U CPU @ 1 60GHz
Programme Language	Python3.9

4.2 Pre-Process

A. Remove the Meaningless Word and Empty Data

In this step, this paper used Python to do the following: 1) Remove the number; 2) Remove punctuation; 3) Remove spaces. 4) Remove the data without the content

B. Stopwords Removal and Text Tokenization

We use jieba as the word splitter. After word tokenization, many meaningless words and symbols remains, which need to be eliminated before topic modeling. We used a stop words list which is obtained by merging several Chinese stop words lists from the Internet to remove these words in this paper. The result of

Table 2. Distribution description of the datasets environment.

Datasets	Positive	Negative	Sum
Train	1419	1508	2923
Test	581	492	1073

the partitioning is shown in the figure below:

如家，一贯的风格，房间也是这样，怎么说，如家的价格还是比较实在，在上海住店，能干净、价格优惠就行了。。。
==========
['家', '一贯', '风格', '房间', '说', '如家', '价格', '比较', '实在', '上海', '住店', '干净', '价格', '优惠', '就行了']

Fig. 3. Result of Tokenization.

4.3   Visualization of the Pre-processing Results

To better understand the distribution of text sentiment in the datasets, we visualize the pre-processing results by using two forms of data visualization: wordcloud and histograms. This allows us to more intuitively and aesthetically grasp the focus of user comments. The visualizations in this paper include histograms of text word frequencies and wordcloud of high-frequency terms. This visualization helps us directly comprehend the distribution of different sentiment categories in the datasets, thereby better understanding the focus and emotional tendencies of the comments.

4.4   Topic Modeling Based on BERTopic

We utilize the pre-trained multilingual model, paraphrase-multilingual-MiniLM-L12-v2, to model our datasets, which is trained in over 50 languages including Chinese. In addition, this section uses the visualization tools for displaying results.

4.5   Sentiment Classification of Tourist Hotel Reviews Based on Gated Recurrent Neural Networks

Word Embedding:
Before training the GRU model, we first perform word embedding on the datasets. This study uses a pre-trained word vector file obtained using the Skip-Gram with Negative Sampling (SGNS) algorithm, with training data from Baidu baike. The model uses word pairs as training samples, with a window size of 5 to consider context words within 5 words of the target word, and word vectors

of dimension 300. In the code, we import this pre-trained word vector file and use it to process the input text.

Evaluating Indicator:

In this study, we focus on classifying hotel review texts, categorizing each text based on the sentiment, either as positive or negative. This is a typical binary classification problem. For such binary classification tasks, accuracy is commonly used as the metric to evaluate model performance. Accuracy is defined as the ratio of correctly classified comments to the total number of comments in the test datasets. Thus, the accuracy measures how well the model can correctly categorize the sentiments of the review texts in the given test set.

$$\frac{TP + TN}{TP + TN + FP + FN} \tag{1}$$

ACC represents the accuracy of classification results; TP is the number of comments correctly classified as positive semantics; FP is the number of comments incorrectly classified as positive semantics; TN is the number of comments correctly classified as negative semantics; and FN is the number of comments incorrectly classified as negative semantics.

Parameter Setting:

In the code, the model parameters are set as follows: The maximum sequence length for review texts is 200, with shorter sequences padded to 200 and longer ones truncated to 200. The vocabulary size is capped at 20,000, with less frequent words marked as unknown to manage data size. The batch size is 32, and the embedding dimension is 300. The GRU network uses 128 hidden units and consists of one GRU layer. The learning rate is 0.007, with a learning rate decay factor (scheduler.gamma) of 0.1 and an adjustment interval (scheduler.step.size) of 5 epochs. The output layer has 2 neurons for binary classification, and the number of training epochs is set to 10.

5   Results

5.1   Results of Pre-Processing

The frequency of pre-processed word is visualized, which is shown in the Fig. 4.

Generate word clouds for two types of sentiment comments, as shown in the Fig. 5.

The word frequency difference between positive and negative comments is shown in Fig. 6.

5.2   Results of GRU

After running the code of GRU and training the network for ten epochs using the training set, the model achieved an accuracy of approximately 0.81 on the test set, with the highest accuracy reaching 0.81. The training results are shown in the following figure.

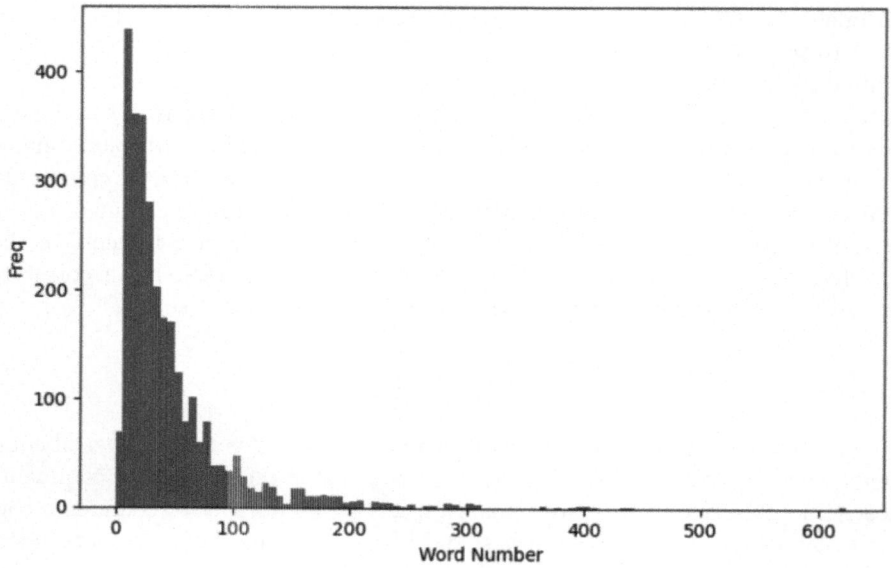

Fig. 4. High-frequency words.

Fig. 5. Wordcloud maps of positive comments and negative comments.

5.3   Text Clustering Results Based on BERTopic

We use BERTopic for topic modeling and the Part of the results are shown in Fig. 9.

We visualized the top5 feature words across 10 topics in Fig. 10 based on the BERTopic.

The keywords included in each cluster are shown in the following table:
Based on the topics displayed, we can conclude that visitors' main concerns regarding hotel accommodation are: Topic0 (Room Service), Topic1 (Travel Convenience), Topic2 (Room Conditions), Topic3 (Service Level), Topic4 (Soundproofing Quality), Topic5 (Cost-Effectiveness), Topic6 (Overall Rating), Topic7 (Newness of Facilities), Topic8 (Advertising), and Topic9 (Network Configuration).

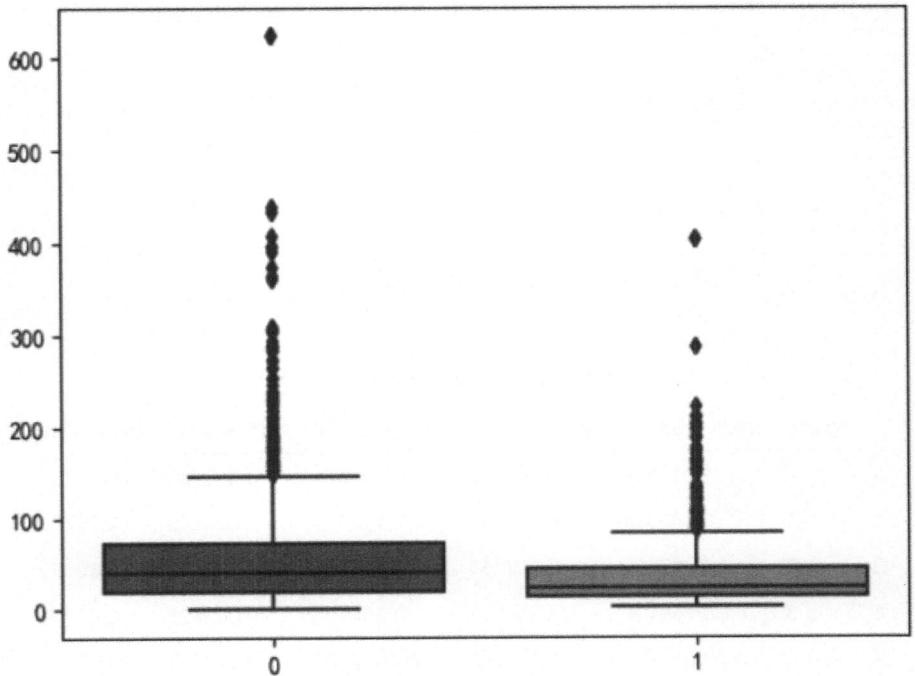

Fig. 6. Word distribution of positive versus negative comments.

Based on the hierarchical clustering visualization shown in the Fig. 11, we can merge similar topics and discard irrelevant ones. This allows us to pinpoint visitors' main concerns about hotels as: Topic1 (Transportation Convenience); Topic2 (Room Cleanliness); Topic3 (Staff Service Quality and Attitude); Topic4 (Noise and Soundproofing); Topic5 (Cost-Effectiveness); Topic6 (Room Decoration and Facility Quality).

5.4   Analysis of Concerns About the Visitor Hotel Choice

1. Transportation Convenience: Keywords for this topic include: convenient, subway station, transportation, subway, train station. This indicates that the proximity of the hotel to transportation hubs like subway stations and train stations is a key factor for visitors when choosing accommodation. The inclusion of keywords like attractions and close suggests that visitors prefer hotels near popular sights and transportation hubs for convenience.
2. Room Cleanliness: This topic includes keywords such as clean, hygiene, and essential. Cleanliness is crucial for maintaining a hotel's reputation and is a significant factor in overall ratings. It is a fundamental requirement for hotel management and can trigger public sentiment issues if neglected. Current issues in hotel hygiene management include: 1) focusing only on visual indicators without considering physical and chemical standards, as many hotels

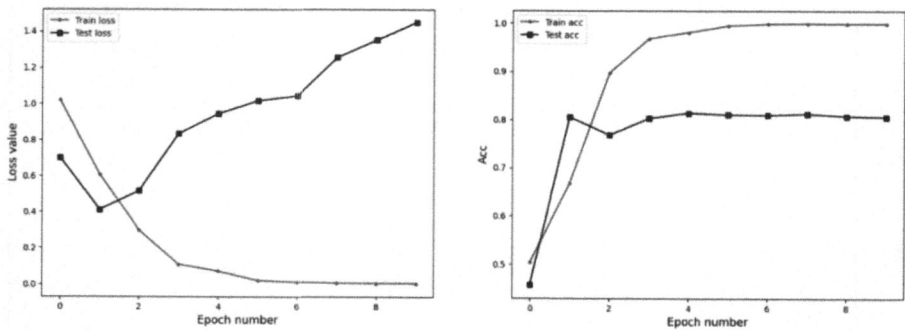

Fig. 7. Results after ten rounds of training.

	epoch	train_loss_all	train_acc_all	test_loss_all	test_acc_all	learn_rate
0	0	1.018248	0.503246	0.700494	0.457596	0.00700
1	1	0.606639	0.666894	0.412073	0.805219	0.00700
2	2	0.297665	0.897164	0.516071	0.767940	0.00700
3	3	0.109126	0.968227	0.834197	0.802423	0.00700
4	4	0.071902	0.980868	0.944708	0.813607	0.00700
5	5	0.020294	0.994534	1.015962	0.809879	0.00007
6	6	0.012705	0.998292	1.043868	0.808947	0.00070
7	7	0.007068	0.998975	1.256795	0.811743	0.00070
8	8	0.005828	0.998633	1.353316	0.807083	0.00070
9	9	0.005699	0.998975	1.451047	0.805219	0.00070

Fig. 8. Ten rounds of training.

Documents and Topics

- 0_酒店_房间_服务
- 1_万便_地铁站_交通
- 2_房间_不错_干净
- 3_热情_熟练_业务
- 4_隔音_太吵_声音
- 5_性价比_价格_不错
- 6_不错_好好_nan
- 7_太老旧_太老_老旧
- 8_copyright1999_广告业务_英才
- 9_宽带_上网_收费
- 10_蚊子_打死_蚊香
- 11_再也不会_再住_感觉
- 12_干净_卫生_主要
- 13_没什么_噪音_隔音

Fig. 9. Review clustering results based on BERTopic.

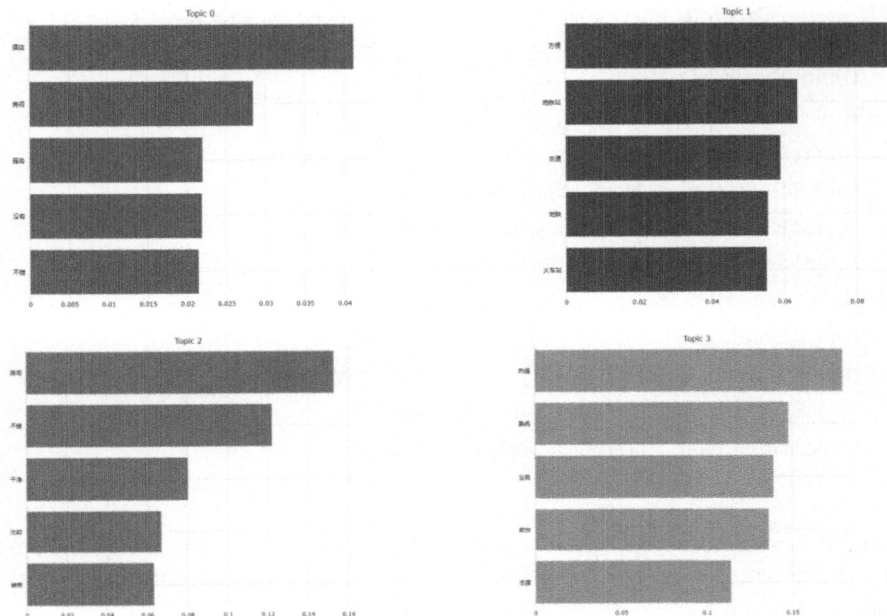

Fig. 10. Subject words of some topics.

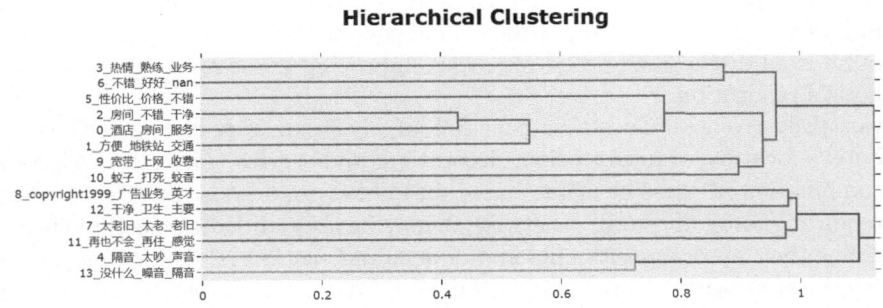

Fig. 11. Subject words of some topics.

rely solely on visual inspections rather than scientific methods; 2) operational procedures may have hygiene management gaps; 3) inadequate air quality management, with many hotels failing to meet air quality standards and having excessive bacteria.

3. Staff Service Ability and Attitude: This topic includes keywords such as enthusiasm, proficiency, business, front desk, attitude, hotel, room, and service. Improving hotel service quality is a key measure for the hotel's continued development and competitive advantage. By enhancing staff training and skills, establishing comprehensive service processes and standards, focusing on customer feedback and complaint handling, innovating service experiences

Table 3. Part of the theme words contained in each topic.

Topic	Words of topic
Topic0	Hotel, rooms, service, none, good
Topic1	Convenient, subway station, transportation, subway, railway station
Topic2	Rooms, nice, clean, compare, decoration
Topic3	Enthusiasm, proficiency, business, reception, attitude
Topic4	Sound insulation, too noisy, sound, room, unbearable
Topic5	Cost performance, price, good, affordable, cheap
Topic6	Good, good, nan, ok, 20
Topic7	Too old, old, facilities
Topic8	Copyright1999, Advertising business, talent, agent, recruitment
Topic9	Broadband, Internet access, charge, Internet, can not

and value-added services, and establishing a customer relationship management system, hotels can improve service quality, enhance customer experience, and gain more customer recognition and support.

4. Room Soundproofing: Soundproofing capability and noise levels are crucial factors for guests when comparing hotels. As people's living standards improve, their expectations for hotels increase. A quiet environment is essential for ensuring guests' rest and comfort. Therefore, effective soundproofing in hotels is particularly important. Good soundproofing can prevent external noise from disturbing guests' rest, thus improving guest satisfaction and the hotel's reputation.

5. Cost-Effectiveness: Guests' satisfaction largely stems from factors such as the hotel's location, transportation, decoration, and service. The price reflects a combination of these elements based on market conditions. Effective pricing requires guests' approval; otherwise, it may be skewed. Hotels need to clearly define their market positioning and understand customers' perceptions of the relationship between price and accommodation experience to set prices that support long-term development.

6. Room Design and Facilities: Exceptional hotel design enhances customer satisfaction by creating a comfortable, safe, and enjoyable environment. Facilities such as internet and projectors significantly improve the guest experience. In a competitive market, well-designed rooms and comprehensive facilities can distinguish a hotel and attract more guests.

According to the analysis presented in the preceding two subsections on the hotel industry in China, I would like to put forward the following views:

1. Cultivating Employees' Service Awareness and Skills: Employees' service awareness is one of the fundamental qualities for enhancing service quality and is a basic guarantee for high service standards. Often, when hotel service quality falls short and guests complain about the service, it's not due to

the employees' lack of service skills or operational proficiency, but rather a deficiency in the essential service awareness and understanding of what "service" truly means and what is required from service personnel. Therefore, it is important to train employees on service awareness, strengthen their skills, and ensure they provide "smile service" to guests.

2. Scientific Site Selection: Site selection should be based on market demand, with a thorough understanding of the target customer group's consumption habits, preferences, and purchasing power, to ensure that the hotel can meet market demands. The principle of convenient transportation is crucial: the hotel should be located in a spot with easy access to public transportation, highways, and airports. The principle of moderate competition should also be considered: avoid direct competition with powerful rivals and ensure the hotel can secure a share of the market.

3. Enhancing Employees' Hygiene Awareness and Professional Ethics: Hotels should improve employees' hygiene awareness and professional ethics. Introduce "laboratory-like" practices in hotel rooms to enhance the physical and chemical hygiene quality. Hotels should not solely rely on periodic inspections by hygiene departments but should invest in their own testing equipment and use simple and effective testing methods. This helps to strictly control hygiene, detect issues promptly, and ensure that rooms with hygiene problems are not provided to guests. Address hygiene blind spots by completing operational procedures.

4. Upgrading Hotel Facilities: Hotel managers should establish a detailed facility inspection plan and strictly implement it. This plan can include weekly, monthly, and quarterly inspection items to ensure the proper operation of hotel facilities. The inspection can be divided into three categories: external facilities, internal facilities, and infrastructure. External facilities include the lobby, parking lot, garden, etc.; internal facilities include guest rooms, restaurants, gyms, etc.; infrastructure includes electricity, water supply, air conditioning, etc. Regular inspections and upgrades of facilities can help promptly identify and resolve issues, avoiding greater losses and inconveniences.

5. Adopting a Reasonable Pricing Strategy: Develop attractive pricing strategies. Guest satisfaction primarily stems from the hotel's location, transportation, decoration, and service. Pricing is a comprehensive reflection of these factors based on market conditions. Effective pricing requires guest approval; otherwise, it may be skewed. When selling externally, hotels should maximize their genuine selling points. Through effective marketing, hotels should convey their strongest selling points to guests.

6   Discussion

The experimental results demonstrate that BERTopic topic clustering can capture key points from tourist reviews, thereby identifying issues in hotel management, highlighting selling points, and addressing tourist concerns, leading to improvements and upgrades for the hotel. Additionally, by employing sentiment analysis based on GRU networks and text topic extraction using BERTopic, it is possible to achieve large-scale monitoring of online public sentiment. This enables scientific management of hotel-related public opinion, facilitating crisis prevention and resolution, which in turn enhances the overall service level of the hotel industry, boosts the city's tourism attractiveness, and avoids large-scale public opinion risks related to hotel services in the tourism sector.

Acknowledgements. This work is supported by the Natural Science Foundation of Shandong Province, China (No.ZR2021MF118, No.ZR2020LZH008, No.ZR2022LZH003), the Key R&D Program of Shandong Province, China (No.2 021SFGC0104, No.2021CXGC010506), Shandong Provincial Project of Graduate Education Quality Improvement, China (No.SDYJG21104, No. SDYJG1917), the OMO Course Group "Advanced Computer Networks" of Shandong Normal University, China, the National Natural Science Foundation of China under Grant (No.62101311) the Shandong Provincial Project of Graduate High Quality Education and Teaching Resources, China (No.SDYKC2022053, No.SDYAL2022060) , and the Undergraduate Teaching Reform Research Project of Shandong Normal University (No.2024ZJ43).

References

1. Tianyu, J., Chen, Z., Fangyuan, L., et al.: Analysis of urban hotel two-factor experience characteristics and their difference mechanism based on Deep Learning. J. Earth Inf. Sci. **26**(02), 499–513 (2024)
2. Xin, L., Ying, W., Xiangbin, Y., et al.: Tourism demand prediction for fine-grained emotion mining of multi-source data. Theory Pract. Syst. Eng. **44**(07), 2293–2308 (2024)
3. Zhou, Q., Jiang, L., Cheng, T.H., et al.: Big data online public opinion of tourism hotels driven by emotional tendency. J. Hunan Univ. Sci. Technol. Nat. Sci. Ed. **35**(04), 67–73 (2020). https://doi.org/10.13582/j.cnki.1672-9102.2020.04.010
4. Subasic, P., Huettner, A.: Affect analysis of text using fuzzy semantic typing. IEEE Trans. Fuzzy Syst. **2**(4), 483–496 (2001)
5. Pang, B., Lee, L., Vaithyanathan, S.:Thumbs up? Sentiment classification using machine learning techniques. In: Empirical Methods in Natural Language Processing, pp. 79–86 (2002)
6. Yang, Z., Qiqi, L., Yuhan, C., et al.: User of overseas shopping APP satisfaction study based on sentiment analysis of online reviews. Data Anal. Knowl. Discov. **2**(11), 19–27 (2018)
7. Kim, Y.:Convolutional neural networks for sentence classification. arXiv preprint arXiv:1408.5882 (2014)
8. Wei, W., Ning, G., Yuting, X., Hongwei, W.: LDA-based analysis of the dynamic evolution of crowdfunding projects. Data Anal. Knowl. Discov. **5**(10), 103–123 (2021)
9. Gu, Y.: Text emotion analysis based on deep learning. Jiangnan University, Jiangsu (2022)
10. Ziyin, W.: Research on Emotion Analysis Based on the Deep Learning Model. Tianjin University of Technology, Tianjin (2022)

11. Yunlong, Y., Jianqiang, S., Guochao, S.: Text sentiment analysis based on gating loop unit and capsule features. Comput. Appl. **40**(09), 2531–2535 (2020)
12. Wang, Z., Wang, H.: Quality evaluation of dissertation based on DBSCAN clustering. In: 2021 11th International Conference on Information Technology in Medicine and Education (ITME) 2021, pp. 460-464. https://doi.org/10.1109/ITME53901. 2021.00098

Privacy-Preserving Covert Channels in VoLTE via Inter-Frame Delay Modulation

Xiaokai Wu[1], Xuan sun[1], Jiaxin Huang[2], Ning Shi[3], and Chen Liang[1]([✉])

[1] Computer School, Beijing Information Science and Technology University,
Beijing 100192, China
20202441@bistu.edu.cn
[2] China Southern Airlines Co., Ltd., Guangzhou 510000, Guangdong, China
huangjiax@csair.com
[3] Hebei Key Laboratory of IoT Blockchain Integration, Shijiazhuang, China
2010014@sjzc.edu.cn

Abstract. The rapid advancement of artificial intelligence and mobile communication technologies has brought unprecedented convenience, but also significant privacy and security concerns, particularly in real-time video communications, such as deepfake. Several researchers have investigated multiple security measures, with Covert Timing Channels (CTCs) emerging as a promising technique for secure information transmission, yet most of them suffer from low transmission efficiency, limiting their real-time responsiveness and network adaptability, making them inadequate for high-capacity covert information transmission. To address this limitation, this paper proposes a novel CTC scheme that modulates the Inter-Frame Delay (IFD) between audio and video frames during VoLTE(Voice over LTE) video calls to transmit covert information. Furthermore, to maintain the undetectability of the channel, a fitting algorithm is used to ensure that the cumulative distribution function (CDF) of the IFD of covert traffic aligns with that of legitimate traffic. The design also incorporates an embedding factor that allows communicating parties to dynamically adjust network parameters, balancing robustness, transmission efficiency, and undetectability. Experimental results demonstrate that, compared to existing methods, the proposed IFDCTC scheme achieves higher transmission efficiency, with a peak transmission rate of up to 30 bps, while maintaining acceptable bit error rates and strong undetectability.

Keywords: Covert Timing Channel · VoLTE · Real-time Video · Inter-Frame Delay

1 Introduction

The swift advancement of mobile communications and artificial intelligence has introduced considerable convenience, yet it has also elicited apprehensions about

F. Zhang et al. (Eds.): AIS&P 2024, LNCS 15399, pp. 74–88, 2025.
https://doi.org/10.1007/978-981-96-1148-5_7

the security and privacy of personal information and sensitive data conveyed through public channels. Deepfake technologies driven by AI, such as Deep-FakeLive [1] , have further heightened concerns, particularly in real-time video communications, where impersonation and data misuse pose significant challenges to privacy and identity verification challenges [2,3,3]. To address these issues, information hiding techniques have emerged as a research focus. These techniques conceal the existence of secret information by embedding it within publicly available carriers, providing a more covert means of protection. Covert channels, as a key method of information hiding, offer a new solution for secure and reliable data transmission in untrusted mobile communication environments, enhancing privacy protection.

In a typical covert channel [4–6], the sender transmits information to the receiver in a concealed manner [7]. Based on their construction methods, covert channels are generally divided into Covert Storage Channels (CSC) [8–10]and Covert Timing Channels(CTC) [11,12]. A csc denotes a technique in which the transmitter inscribes data in designated storage places (such as memory cells, resource states, or network packets), and the recipient acquires the information by accessing these sites [13]. In contrast, a ctc involves the sender influencing system events (such as performance or behavior), while the receiver interprets the transmitted information by observing the sequence, interval, or frequency of these events [14–16].

Currently, LongTerm Evolution (LTE) has become the dominant mobile access technology globally, with more than 3 billion users, and it is expected to maintain this dominance in the coming years. VoLTE as the transition path for voice services over all LTE networks, provides high-quality voice and video call experiences, making it a crucial component of next-generation communication services. Although VoLTE delivers excellent communication services, it also faces certain privacy and security threats. Yu et al. [17] have confirmed that there are vulnerabilities in LTE networks, which allows attackers to disable information protection mechanisms in both LTE and IP Multimedia Subsystem (IMS). Moreover, with the rise of AI-driven deepfake technologies, VoLTE video calls are increasingly vulnerable to impersonation attacks, where attackers can generate highly realistic fake videos to deceive the receiving party. Hence, it is necessary to establish effective covert channels within VoLTE to ensure the security of private data and to enhance identity verification during real-time communications.

In the VoLTE context, several effective schemes have been proposed for the construction of covert channels. The primary methods include packet reordering and deliberate packet dropping. However, these approaches still face significant challenges. For example, packet reordering may introduce substantial latency and is prone to detection, while deliberate packet dropping can negatively impact voice and video quality, thus degrading the user experience [15,18]. Furthermore, these methods often exhibit low transmission efficiency, which makes them less suitable for large-scale privacy data transmission [16,19]. To address these issues, further optimization of covert channels in the VoLTE environment remains a pressing research direction. Consequently, the main motivation of this paper is to construct a covert timing channel with high transmission efficiency, ensuring the reliable transmission of large-scale privacy data.

The study designs a novel covert timing channel for VoLTE based on IFD between audio and video frames, referred to as IFDCTC. Secret information is modulated into the IFD during video calls between audio and video frames. The main contributions of this research are as follows:

1. A novel covert timing channel algorithm is proposed for VoLTE, where the sender transmits secret information by modifying the IFD, and only the authorized receiver is able to decode the information.
2. The proposed algorithm ensures undetectability, meaning the IFD distribution of the covert channel remains consistent with that of the overt channel, allowing information to be securely transmitted without detection.
3. Experimental results demonstrate that, compared to other schemes, IFD-CTC offers higher transmission efficiency while maintaining undetectability and robustness, with a peak transmission rate of up to 30bps under optimal conditions.

The subsequent sections of this work are structured as follows: In Sect. 2, we examine the pertinent literature on established covert channels. Section 3 presents the preliminaries, encompassing VoLTE video calls and the inter-frame delay of voice and video frames. In Sect. 4, we present the system model design for the IFDCTC and discuss its performance metrics, including undetectability, robustness, and transmission efficiency. Section 5 demonstrates the construction of the covert channel. In Sect. 6, we provide experimental results and analysis. Finally, in Sect. 7, we conclude the paper and discuss future research directions.

2   Related Works

As computer network technology progresses, covert channels have attracted significant attention as an efficient method for sending concealed information while circumventing security protocols [20–22]. Early research primarily focused on traditional wired networks, particularly in Ethernet environments.

However, with the rapid proliferation of wireless networks and the increasing security threats associated with them, the construction of covert channels in wireless network environments has emerged as a new research focus [23,24]. In IEEE 802.11 networks, covert channel construction typically leverages the protocol's unique characteristics and mechanisms, such as the backoff mechanism and Beacon frames [25,26]. Teca et al. [27] proposed the StegoBackoff scheme, which manipulates the 802.11 backoff mechanism by encoding covert information into the backoff time before packet transmission, thereby enabling covert communication.

Regarding the VoLTE mobile network scenario, researchers have proposed various covert channel implementation schemes [19,20]. Liang et al. [20] introduced an active packet dropping covert timing channel, which effectively reduces the interference of network noise through multilevel verification and error correction strategies, thereby achieving a low bit error rate. Zhang et al. [19] proposed a covert timing channel based on packet reordering, where the covert information is transmitted by modulating the number of RTP packets between RTCP packets. Li et al. [18] utilized parity concatenated coding and codeword verification to construct a robust active packet dropping covert timing channel.

Despite the efficacy of numerous existing systems in undetectability and robustness, they typically exhibit low transmission efficiency, hence constraining their use for effective private data transmission. Drawing inspiration from the existing CTC construction schemes, this paper proposes IFDCTC utilizing the in IFD between audio and video frames. Since the audio and video data streams originate from different sources, their frames are not completely synchronized, allowing us to exploit this characteristic for modulation and embedding covert information. Furthermore, thanks to the various buffering mechanisms employed by both communication terminal, the modulation process has minimal impact on the overall communication quality, resulting in high transmission efficiency and strong resistance to detection.

3   Preliminary

3.1   VoLTE Video Calls

VoLTE is a communication technology based on the 4G LTE network, designed to provide high-quality, low-latency voice and video calling services. By directly transmitting voice and video data packets over the LTE network, VoLTE achieves the convergence of voice and data services. Unlike traditional circuit-switched voice communication, VoLTE utilizes an IP Multimedia Subsystem (IMS) architecture and communicates through packet switching, significantly enhancing the quality of voice and video calls.

In VoLTE video calls, the user equipment is registered through the IMS, establishes session connections, and negotiates media parameters. Communication data is transmitted through the Realtime Transport Protocol (RTP) and the Realtime Transport Control Protocol (RTCP). Voice data is processed using the Adaptive MultiRate Wideband (AMR-WB) codec, while video data is encoded using the H.264 codec. To ensure the transmission quality of critical data, voice packets are assigned a higher priority than video packets, as voice communication is highly sensitive to latency. Therefore, compared to voice packets, video packets are more suitable as carriers for transmitting covert information.

3.2   Inter-Frame Delay for Video and Audio

The analysis of captured VoLTE video streams reveals three primary data flows in VoLTE video calls: the first type comprises SIP signaling packets used for session negotiation; the second type consists of voice data streams; and the third type involves video data streams. SIP signaling packets are transmitted randomly and occur infrequently, making them unsuitable as carriers for covert channels. Voice data streams are composed of voice frames, each frame containing a single voice packet. Video data streams, due to the large amount of information contained in a single video frame, are divided into multiple video packets for transmission. Depending on the hardware of the device, the sampling rate of video frames varies, with a maximum of 30 frames per second.

In video calls, there is typically a delay between audio and video streams, which prevents perfect synchronization. It is generally accepted that a delay ranging from -100ms to 50ms is imperceptible to humans. In VoLTE, audio and video stream synchronization is accomplished via timestamps, with data packets corresponding to the same video frame sharing identical timestamps. Using these timestamps and the sampling frequency, the playback timeline can be calculated for audio and video. Taking audio frames as a reference, the closest video frame in terms of playback time is identified, and the IFD is computed. IFD is calcutated by Eq. 1. $R_{video/audio}$ where represents the timestamp of the video or audio frame and $f_{video/audio}$ denotes the frequency of the video or audio frame.

$$\text{IFD}_{v-a} = \min\left(\left|\frac{R_{\text{video}} - R_{\text{video\_start}}}{f_{\text{video}}} - \frac{R_{\text{audio}} - R_{\text{audio\_start}}}{f_{\text{audio}}}\right|\right) \quad (1)$$

As shown in Fig. 1, the IFD distribution approximates a normal distribution and exhibits certain redundancy. Taking advantage of this redundancy, we can subtly adjust the delay between video and audio frames to covertly transmit information.

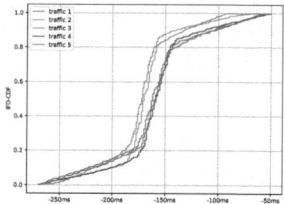

Fig. 1. Cumulative distribution function of IFD for five sets of overt traffic. The IFD distribution approximates a normal distribution, with a range between $-200\,\text{ms}$ and $-100\,\text{ms}$, indicating that audio packets have a higher priority than video packets.

The cumulative distribution of IFD is shown in Fig. 1, where it can be observed that it exhibits an approximately normal distribution, ranging between -200 ms and -100 ms. This indicates that the audio stream is transmitted ahead of the video stream. Additionally, there is a certain amount of redundancy within the IFD. We can exploit this redundancy by finetuning the delay between video and audio frames to construct the IFDCTC.

4   System Design

4.1   System Model

In the IFDCTC system model, there are three roles: sender, receiver, and monitor. The public channel used by the CTC is the data stream generated during VoLTE video calls. The overall system model of the CTC is illustrated in Fig. 2. Before the transmission of covert information, the sender and receiver share a pre-agreed seed for generating the random sequence and a symbol mapping rule. First, the sender encodes the covert information into different symbols, using the random sequence as an index to locate the video frame packets to be modified. By advancing or delaying the IFD of the legitimate audio-video traffic, the symbols are modulated and covertly transmitted to the receiver via the LTE network. The receiver captures the modulated IFD, demodulates it into symbols based on the random sequence, and then decodes the covert information according to the symbol mapping rule. The sender and receiver are defined as the two endpoints of covert communication. The covert information, represented as $\{b1, b2, b3, \dots\}$, is encoded into symbols $\{s1, s2, s3, \dots\}$. Combined with the random sequence , the legitimate VoLTE IFD $\{t1, t2, t3, \dots\}$ are modulated into $\{t'1, t'2, t'3, \dots\}$.The main notations and symbols are shown in Table 1.

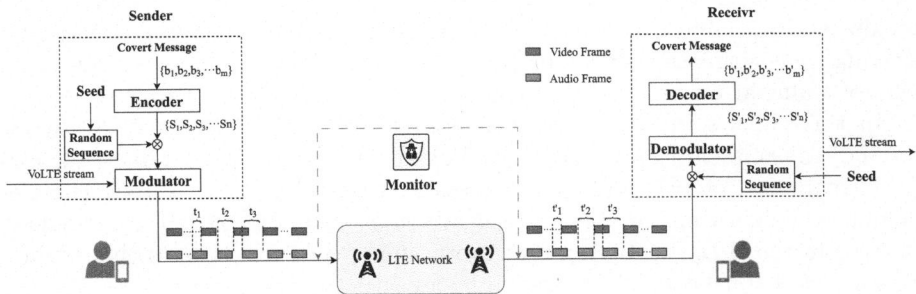

Fig. 2. System Model of IFDCTC in VoLTE Video Calls: A Covert Timing Channel Framework Involving Sender , Receiver , and Monitor. The Sender Encodes Covert Information into Symbols, Utilizes a Random Sequence to Identify Specific Video Frames for Adjustment of IFD, and Transmits the Modulated Traffic through the LTE Network to the Receiver. The Receiver, Using the Shared Random Sequence, Demodulates and Decodes the Covert Information, While the Monitor Observes the Transmission.

4.2   Threat Model

This paper's threat model presupposes that the intruder can either be an intrusion detection system or a covert timing channel disruptor; either way, they can observe or alter the communications between Sender and Receiver. Both official

Table 1. The main notations and symbols

Notation	Description
b_m	The m-th bit of the covert message to be transmitted
b'_m	The m-th bit of the covert message decoded at the receiver
S	The bit sequence of the covert message
S_n	The n-th code symbol generated by the encoder
S'_n	The n-th code symbol decoded at the receiver
F	The original frame sequence (including video and audio frames)
F'	The modulated frame sequence
t_i	The i-th original Inter-Frame Delay (IFD)
t'_i	The i-th modulated Inter-Frame Delay (IFD)
$M(\cdot)$	The mapping function, which maps bit b_i to a corresponding IFD range
P_e	Bit error rate (BER) or the probability of error
T_r	Transmission rate, calculated as $T_r = F_v \times \alpha$
F_v	Frame rate of the video
α	Embedding factor for the covert transmission

and secret communications are accessible to the enemy. Covert traffic cannot coexist with genuine traffic in the transmission channel since it is created by altering the legitimate traffic. The transmission channel can only ever carry one kind of traffic at a time, and that can only be legitimate or covert.

In this paper's threat model, it is assumed that the adversary can monitor the transmission characteristics of VoLTE audio and video frames, such as the distribution of IFD and may possess knowledge of covert timing channel modulation algorithms. However, if the adversary is unable to detect anomalous changes in the IFD, identifying the covert timing channel becomes challenging. The covert channel could be disrupted by an active attacker introducing random interference into the channel. An opponent may try to disrupt legal communication for a short period of time if they believe a covert channel is there. While these measures can influence legal transmission, they won't stop the embedding of secret data. Because of this, the covert channel needs to be very resilient to network noise, whether it's caused by random oscillations in the network or is deliberately injected by an attacker trying to derail it.

4.3 Design Criteria

Based on the model and assumptions presented above, we introduce the design objectives of the proposed IFDCTC scheme.

Undetectability: The primary objective of covert communication is to achieve undetectability, meaning that a passive adversary is unable to detect the existence of the covert channel [28,29]. Currently, no CTC detection algorithm is universally applicable, with most detection approaches relying on statistical analysis to determine the presence of a covert channel. Therefore, we have designed the IFD modulation algorithm to ensure that the distribution of modulated IFDs aligns with that of the overt traffic, thereby enhancing the covert channel's resistance to detection. Additionally, in the experimental section, we compare the cumulative distribution functions (CDF) of IFDs and evaluate the created IFDCTC's undetectability using the Kolmogorov-Smirnov (KS) test and the Kullback-Leibler Divergence (KL-D) test.The KS test compares the cumulative distribution functions of two samples to assess if they come from the same distribution, which helps evaluate the similarity between covert and overt IFDs. The KL-D measures the difference between two probability distributions, quantifying how covert traffic deviates from overt traffic.

Robustness: Due to the potential unreliability of covert channels during transmission, there may be packet loss or abnormal delays, which can disrupt the inter-frame delay of audio and video frames, leading to errors in the information extracted by the receiver. Therefore, the covert channel must exhibit sufficient robustness. Specifically, robustness can be measured by the bit error rate (BER), which ensures that under a given robustness requirement $\epsilon \in R^+$,the decoding error rate achieves $P_e \leq \epsilon$. The bit error rate for this covert channel can be derived from Eq. 2.

$$P_e = \frac{1}{N}\sum_{i=1}^{N}\delta(IFD_i, G(IFD_i)) \tag{2}$$

$\delta(IFD_i, G(IFD_i)$ is an indicator function, where $\delta = 1$, if IFD_i is outside the mapping range $G(IFD_i)$, indicating a decoding error, and $\delta = 0$ ifIFD_i falls within the mapping range, indicating correct decoding.

Transmission Efficiency: The transmission rate of the covert channel constructed in this scheme is determined by the frame rate of the VoLTE video call and the embedding factor. The communi-cating parties can select an appropriate embedding factor based on different network conditions to balance undetectability and transmission efficiency. The specific transmission efficiency Tr(bps) can be derived from the following Eq. 3:

$$T_r = F_v \times \alpha \tag{3}$$

5   Covert Timing Channel Construction

This section provides a detailed explanation of the construction process for a CTC, covering the information modulation and extraction procedures within the system model. To achieve an effective CTC, the sender and receiver must establish an agreement over the public channel. This agreement includes selecting an encoding scheme that maps secret information bits to encoded symbols along with related parameters, and generating a seed for the random sequence used for modulation. The encoded symbols are modulated into IFD of the audio and video frames to enable covert communication.

5.1 Symbolic Mapping

By utilizing IFD between audio and video frames in VoLTE streams, both the sender and receiver observe similar statistical distribution characteristics. This enables the construction of symbol mapping rules based on the CDF of the IFD. Specifically, encoding symbols can be mapped one-to-one with specific delay intervals. For instance, a simple mapping rule might assign 0 to the interval 0 $\tilde{0}.5$ and 1 to the interval 0.5 $\tilde{1}$. Depending on the network conditions encountered by both communication parties, the mapping rules can be further optimized to balance between undetectability and transmission efficiency.

5.2 Modulation

In this suggested approach, the confidential information is initially encoded into symbols, which are subsequently modulated into the IFD of the audio and video frames. Specifically, each encoded symbol s_i is mapped to an IFD t_i through a reversible mapping function $M(\cdot)$ This mapping ensures that each symbol s_i corresponds uniquely to a specific IFD t_i, allowing the receiver to recover the information using the inverse of the mapping function $M^{-1}(\cdot)$. The process is expressed by the following Eq. 4.

$$t_i := M(s_i), i = 1, 2, 3, 4, \ldots \tag{4}$$

At the receiving end, the received IFD t_i' is demodulated through the inverse mapping function $M^{-1}(\cdot)$, recovering the original symbol s_i':

$$s_i' := M^{-1}(t_i'), i = 1, 2, 3, 4, \ldots \tag{5}$$

This modulation process ensures the accurate transmission of information and the correct recovery of the original data at the receiver's end.

In the modulation process of IFD, as illustrated in Fig. 3, the secret information is first encoded into symbols and then converted into IFD using predefined symbol mapping rules. The sender generates a random sequence based on a pre-established seed to determine the specific audio-video frames that need modulation. This random selection enhances the stealthiness of the process, making it more difficult for an attacker to detect a pattern in the changes of IFD. Next, the IFD between the selected video frame and its corresponding audio frame, denoted as t_1, is calculated according to Eq. 1. A bit is then extracted from the secret information and mapped to the corresponding IFD range based on the predefined mapping rules. For example, a bit '0' may be mapped to the interval [0, 0.5], while a bit '1' may be mapped to [0.5, 1], using the mapping function $M(\cdot)$. If the original IFD t_1 does not fall within the target range of the symbol mapping, a fitting algorithm is applied to adjust it to the corresponding symbol

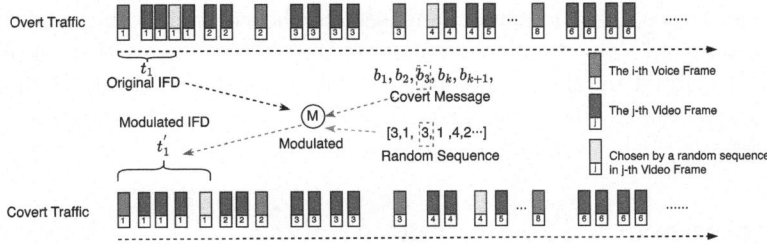

Fig. 3. The IFDCTC Modulation Process: A Step-by-Step Description from Secret Information Encoding to IFD Adjustment. This process involves encoding secret information into symbols, generating a random sequence based on a predefined seed, selecting specific frames according to the random sequence, and calculating the original IFD.

range, resulting in a modulated delay t_1'. Finally, the video frame is transmitted with the new IFD t_1' .Algorithm 1 details the steps of the modulation process.

Algorithm 1. Frame IPD Modulation Algorithm

1: **Input:** Secret message S, Frame sequence F, Random seed *seed*, Mapping function $M(\cdot)$
2: **Output:** Modulated frame sequence F'
3: RandomSeq $\leftarrow$ GenerateRandomSequence(seed, F)
4: $F' \leftarrow F$
5: **for** $i = 1$ to length(S) **do**
6:     FrameIndex $\leftarrow$ RandomSeq[i]
7:     $b_i \leftarrow S[i]$
8:     $t_i \leftarrow$ CalculateIFD(F[FrameIndex])
9:     TargetRange $\leftarrow M(b_i)$
10:     **if** t_i not in TargetRange **then**
11:         $t_i' \leftarrow$ AdjustIFD(t_i, TargetRange)
12:     **else**
13:         $t_i' \leftarrow t_i$
14:     **end if**
15:     $F'[FrameIndex] \leftarrow$ ModulateFrame(F[FrameIndex], t_i)
16: **end for**
17: **return** F'

5.3 Decoding

The demodulation process at the receiver's side is the inverse of the modulation operation. After buffering the received audio and video frames, the receiver calculates the IFD based on the incoming data and reconstructs the encoded symbols according to the predefined mapping rules, ultimately extracting the secret information. Initially, the receiver generates a random sequence using the same random seed as the sender and selects the corresponding video frames from the received data to calculate their IFD t_1'. The receiver then applies the pre-agreed symbol mapping rules to map the received IFD back to the corresponding symbols. By iterating this process for each frame, the receiver can fully demodulate the IFD sequence and recover the complete secret message.

6   Experimental Results and Analysis

6.1   Experimental Setup

Our proposed scheme can be successfully implemented on different models and versions of Android smartphones, provided that the devices and systems support VoLTE video calls. For our experiments, we selected two Xiaomi 8 smartphones running Android version 12. In these experiments, both the sender and receiver captured the VoLTE traffic. Additionally, to balance the undetectability and throughput of the covert channel, we introduced the parameter α, which represents the embedding rate of covert information per second. The value of α can be negotiated by the communicating parties based on network conditions and specific requirements.

6.2   Undetectability

We analyze the undetectability of the constructed covert channel from three experimental perspectives. First, we conduct a statistical test to assess the undetectability of the covert channel by calculating the cumulative distribution function (CDF) of the IFD for both overt and covert traffic. Second, we employ two standard statistical tests, the Kolmogorov-Smirnov (KS) test and the Kullback-Leibler Divergence (KLD) test, to visualize and further validate the undetectability of the channel [30,31].

The CDF of the covert traffic is shown in Fig. 4. The CDFs of the four embedding factors $(1/4, 1/2, 3/4, 1)$ all fit well with the CDF of legitimate traffic. In comparison, the CDF difference between legitimate traffic and covert traffic in Fig. 4 is significantly smaller than the CDF differences observed in overt traffic shown in Fig. 3. Additionally, it can be seen from the figure that the embedding factor α has a minimal impact on the CDF.

The K-S test results are presented in Fig. 5. Based on the p-value, the test determines whether the two sample distributions differ; when $p > 0.05$, it is concluded that the two samples follow the same distribution. From the KS test results of the covert traffic, it is observed that as the α decreases, the p-value approaches 1. Therefore, in environments with higher network fluctuations, reducing the embedding factor α can lower transmission efficiency but ensures enhanced resistance to detection.

The KLD test results are shown in Table 2. The smaller the KLD value, the smaller the difference between the two probability distributions. Conversely, the larger the KLD value, the greater the distributional difference. As observed

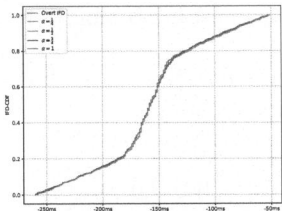

Fig. 4. Cumulative Distribution Function (CDF) comparison between covert traffic and legitimate traffic under different embedding factors. The CDFs for four distinct embedding factors (1/4,1/2,3/4, 1) demonstrate a close alignment with the CDF of legitimate traffic, illustrating minimal deviation.

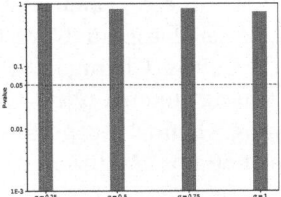

Fig. 5. K-S Tests for IFD Under Different α. As the α decreases, the P-value increases, indicating that smaller embedding factors improve the resistance to detection, enhancing the covert channel undetectablity

from the table, even when the embedding factor equals 1, the KLD value remains well below 0.1, indicating that the proposed FrameIPD scheme maintains strong undetectability even at the highest transmission efficiency.

Table 2. KLD Tests for IFD Under Different α.

α	KLD-tests
$\frac{1}{4}$	4.35×10^{-3}
$\frac{1}{2}$	8.70×10^{-3}
$\frac{3}{4}$	9.64×10^{-3}
1	1.82×10^{-2}

6.3   Transmission Rate

The transmission rate of our proposed covert channel scheme is directly related to the frame rate during VoLTE video calls and the covert information embedding factor α. When $\alpha = 1$, the optimal transmission rate of the covert channel is achieved, where secret information is embedded in the IFD of all video frames. The capacity of the IFDCTC ranges from 1 to 30 bits per second.

Table 3. Transmission Rate comparison of CTC Schemes

CTC Scheme	Transmission Rate (bps)
RPDCTC	1.0–2.0
SPCTC	0.8–3.0
NoP-GCTC	1.0–8.0
IFDCTC	1.0–30.0

Various schemes for constructing covert timing channels in VoLTE scenarios have been proposed. A comparison of our constructed audio-video frame-based IPD covert channel with these schemes is shown in Table 3. These schemes include the Robust Packet-Dropout Covert Channel (RPDCTC) [18], the Adjusting Silence Periods Covert Channel (SPCTC) [16], and the packet rearrangement-based covert timing channel (NoP-GCTC) [19]. Our audio-video frame IPD-based covert timing channel outperforms most of these schemes, achieving a transmission rate of 30 bps. Additionally, the bit error rate (BER) of decoding on the receiver side is less than 10%, which is acceptable and suitable for transmitting large amounts of secret information.

7   Conclusion and Future Work

AI-driven deepfake technologies have amplified these concerns, particularly in real-time video communications, making privacy protection and identity verification crucial. Covert channels provide a secure method for transmitting private information over wireless networks, offering effective solutions for real-time identity verification and privacy-preserving communication.This work presents a VoLTE timed covert channel utilizing IFD of audio-video frames, wherein covert information is encoded into the IFDs of audio and video streams. This establishes a covert channel characterized by elevated transmission efficiency and resilience against detection. To guarantee the covert channel's undetectability, a model fitting technique specifically designed for the traffic characteristics of VoLTE audio-video frames is utilized, rendering the distribution characteristics of covert traffic essentially indistinguishable from those of overt traffic. An embedding factor is established, enabling both communicating parties to negotiate parameters according to network conditions to optimize robustness, transmission efficiency, and undetectability of the covert channel. Comprehensive experimental findings validate the efficacy of our CTC system regarding both undetectability and transmission efficiency.

However, there is still considerable room for improvement in the robustness of our scheme. Under adverse network conditions with significant fluctuations, the receiver may experience a higher bit error rate. In the future, we plan to introduce reliability coding schemes, such as Gray code and fountain codes, to enhance the robustness of the system and improve the usability of the covert channel in extreme environments.

Acknowledgments. This research was supported by R&D Program of Beijing Municipal Education Commission (KM202311232013)

References

1. Yin, Q., Lu, W., Li, B., Huang, J.: Dynamic difference learning with spatiotemporal correlation for deepfake video detection. IEEE Trans. Inf. Forensics Secur. **18**, 4046–4058 (2023)
2. Shang, J., Wu, J.: Protecting real-time video chat against fake facial videos generated by face reenactment. In: 2020 IEEE 40th International Conference on Distributed Computing Systems (ICDCS), pp. 689–699 (2020)
3. Mittal, G., Hegde, C., Memon, N.: GOTCHA: real-time video deepfake detection via challenge-response. In: 2024 IEEE 9th European Symposium on Security and Privacy, pp. 1–20. IEEE Computer Society, Los Alamitos, CA, USA (2024)
4. Zhang, T., Li, B., Zhu, Y., Han, T., Wu, Q.: Covert channels in blockchain and blockchain based covert communication: overview, state-of-the-art, and future directions. Comput. Commun. **205**, 136–146 (2023)
5. Caviglione, L., Mazurczyk, W.: You can't do that on protocols anymore: analysis of covert channels in IETF standards. IEEE Netw. **38**(5), 255–263 (2024)
6. Żórawski, P., Caviglione, L., Mazurczyk, W.: A long-term perspective of the internet susceptibility to covert channels. IEEE Commun. Mag. **61**(10), 171–177 (2023)
7. Costa, G., Pinelli, F., Soderi, S., Tolomei, G.: Turning federated learning systems into covert channels. IEEE Access **10**, 130642–130656 (2022)
8. Bedi, P., Jindal, V., Dua, A.: SPYIPv6: locating covert data in one or a combination of IPv6 header field(s). IEEE Access **11**, 103486–103501 (2023)
9. Dua, A., Jindal, V., Bedi, P.: Covert communication using address resolution protocol broadcast request messages. In: 2021 9th International Conference on Reliability, Infocom Technologies and Optimization (Trends and Future Directions) (ICRITO), pp. 1–6 (2021)
10. Teca, G., Natkaniec, M.: A novel covert channel for IEEE 802.11 networks utilizing MAC address randomization. Appl. Sci. **13**, 8000 (2023)
11. Liang, C., Tan, Y.A., Zhang, X., Wang, X., Zheng, J., Zhang, Q.: Building packet length covert channel over mobile VoIP traffics. J. Netw. Comput. Appl. **118**, 144–153 (2018)
12. Tan, Y.A., Xu, X., Liang, C., Zhang, X., Zhang, Q., Li, Y.: An end-to-end covert channel via packet dropout for mobile networks. Int. J. Distrib. Sens. Netw. **14**(5), 1550147718779568 (2018)
13. Luo, X., Zhang, P., Zhang, M., Li, H., Cheng, Q.: A novel covert communication method based on bitcoin transaction. IEEE Trans. Industr. Inf. **18**(4), 2830–2839 (2022)
14. Tan, Y.A., Zhang, X., Sharif, K., Liang, C., Zhang, Q., Li, Y.: Covert timing channels for IoT over mobile networks. IEEE Wireless Commun. **25**(6), 38–44 (2018)
15. Yuanzhang, L., Junli, L., Xinting, X., Xiaosong, Z., Li, Z., Quanxin, Z.: A robust packet-dropping covert channel for mobile intelligent terminals. Int. J. Intell. Syst. **37**(10), 6928–6950 (2022)
16. Zhang, X., Tan, Y.A., Liang, C., Li, Y., Li, J.: A covert channel over volte via adjusting silence periods. IEEE Access **6**, 9292–9302 (2018)

17. Yu, C., Chen, S., Wei, Z., Wang, F.: Toward a truly secure telecom network: analyzing and exploiting vulnerable security configurations/ implementations in commercial LTE/IMS networks. IEEE Trans. Dependable Secure Comput. **21**(4), 3048–3064 (2024)
18. Li, Y., Zhang, X., Xu, X., Tan, Y.A.: A robust packet-dropout covert channel over wireless networks. IEEE Wireless Commun. **27**(3), 60–65 (2020)
19. Zhang, X., Liang, C., Zhang, Q., Li, Y., Zheng, J., Tan, Y.A.: Building covert timing channels by packet rearrangement over mobile networks. Inf. Sci. **445-446**, 66–78 (2018)
20. Liang, C., Baker, T., Li, Y., Nawaz, R., Tan, Y.A.: Building covert timing channel of the IoT-enabled MTS based on multi-stage verification. IEEE Trans. Intell. Transp. Syst. **24**(2), 2578–2595 (2023)
21. Guri, M.: Air-gap electromagnetic covert channel. IEEE Trans. Dependable Secure Comput. **21**(4), 2127–2144 (2024)
22. Schmidbauer, T., Keller, J., Wendzel, S.: Challenging channels: encrypted covert channels within challenge-response authentication. In: Proceedings of the 17th International Conference on Availability, Reliability and Security. ARES '22, Association for Computing Machinery, New York, NY, USA (2022)
23. Liang, C., Wang, X., Zhang, X., Zhang, Y., Sharif, K., Tan, Y.A.: A payload-dependent packet rearranging covert channel for mobile VoIP traffic. Inf. Sci. **465**, 162–173 (2018)
24. Son, S., Kwon, D., Lee, S., Jeon, Y., Park, Y.: A robust covert channel with self-bit recovery for IEEE 802.11 networks. IEEE Internet Things J. **11**(16), 27356–27368 (2024)
25. Sawicki, K., Bieszczad, G., Piotrowski, Z.: StegoFrameOrder-MAC layer covert network channel for wireless IEEE 802.11 Networks. Sensors **21**(18) (2021)
26. Seong, H., Kim, I., Jeon, Y., Oh, M.K., Lee, S., Choi, D.: Practical covert wireless unidirectional communication in IEEE 802.11 environment. IEEE Internet Things J. **10**(2), 1499–1516 (2023)
27. Teca, G., Natkaniec, M.: StegoBackoff: creating a covert channel in smart grids using the backoff procedure of IEEE 802.11 networks. Energies **17**(3) (2024)
28. Chen, Z., Zhu, L., Jiang, P., Zhang, C., Gao, F., Guo, F.: Exploring unobservable blockchain-based covert channel for censorship-resistant systems. IEEE Trans. Inf. Forensics Secur. **19**, 3380–3394 (2024)
29. Elsadig, M.A., Gafar, A.: Covert channel detection: machine learning approaches. IEEE Access **10**, 38391–38405 (2022)
30. Iv, J.K.H., Georgiou, M., Malozemoff, A.J., Shrimpton, T.: Security foundations for application-based covert communication channels. In: 2022 IEEE Symposium on Security and Privacy (SP), pp. 1971–1986 (2022)
31. Liang, Q., Shi, N., Tan, Y.A., Li, C., Liang, C.: A stealthy communication model with blockchain smart contract for bidding systems. Electronics **13**(13), 2523 (2024)

Enhancing Adversarial Robustness in Object Detection via Multi-task Learning and Class-Aware Adversarial Training

Fan Qin[1], Nan Ji[2], Zeping Ye[3], Ning Shi[4(✉)], and Yu-an Tan[1,5]

[1] School of Cyberspace Science and Technology, Beijing Institute of Technology, Beijing, China
fanqin@bit.edu.cn
[2] Qian Xuesen Laboratory of Space Technology, China Academy of Space Technology, Beijing, China
[3] China Southern Airlines Co., Ltd, 510000 Guangzhou, Guangdong, China
[4] College of Science, Shijiazhuang University, Shijiazhuang, China
shining@sjzc.edu.cn
[5] Shandong Provincial Key Laboratory of Energy Industry Internet Big Data Technology, 250003 Jinan, Shandong, China

Abstract. Object detection plays a critical role in numerous applications, including security surveillance and autonomous driving, where system reliability is crucial. However, modern object detectors, which are based on deep learning, are vulnerable to adversarial attacks that can severely degrade their performance. This paper proposes three key strategies to enhance the robustness of object detection models: (1) a confidence loss mechanism in multi-task learning settings to strengthen overall model robustness, (2) an optimized adversarial training method using fast adversarial training technique to enhance training efficiency, and (3) a class-wise adversarial training approach to ensure robustness is balanced across object classes, preventing weaker classes from being disproportionately affected. Extensive experiments and multiple ablation studies on the PASCAL-VOC and MS-COCO datasets confirm that these strategies significantly improve model robustness.

Keywords: Object Detection · Adversarial Attacks · Multi-Task Learning · YOLOv5

1 Introduction

Object detection is a foundational technology for a wide range of applications, such as autonomous driving [21,23], smart retail systems [27] and robotic automation [9]. The success and reliability of these systems hinge on their ability to accurately identify and localize objects within images. While significant advancements have been made in object detection, particularly with deep learning techniques, the robustness of these models remains a critical concern [2,14,25].

F. Zhang et al. (Eds.): AIS&P 2024, LNCS 15399, pp. 89–104, 2025.
https://doi.org/10.1007/978-981-96-1148-5_8

Deep learning-based object detection models exhibit inherent vulnerabilities, as first demonstrated by Szegedy et al. [22], who showed that these models are susceptible to adversarial examples. Since this discovery, several attack methods have been developed, including the Fast Gradient Sign Method (FGSM) [7], DeepFool [15], and Projected Gradient Descent (PGD) [13]. These methods collectively underscore the susceptibility of deep learning classification models to adversarial manipulations.

To mitigate adversarial attacks, researchers have developed various strategies to enhance the robustness of object detection models [1,5,7,8,16,22,24,26]. Among these, adversarial training has proven to be one of the most effective defense mechanisms. This method trains models using adversarial examples, allowing them to better withstand adversarial perturbations during inference. Object detection is typically divided into two tasks: classification and localization [6,10,18,18]. Classification tasks are optimized using classification loss, whereas localization tasks are refined through localization-specific loss functions. Many adversarial training methods for object detection are often time-consuming and result in models with inconsistent robustness across different object classes. Additionally, these methods frequently overlook the significance of confidence loss in improving model robustness.

In the proposed methodology, we decompose the localization loss into two distinct components: box loss and confidence loss, with the latter playing a pivotal role in improving model robustness. To reduce the high computational overhead of traditional adversarial training, we leverage fast adversarial training techniques [20]. Additionally, we incorporate Class-wise Adversarial Training (CWAT) [1] to ensure balanced robustness across various object categories. Through extensive experiments on the PASCAL VOC [4] and MS-COCO [11] datasets, we demonstrate the effectiveness of our approach, achieving notable improvements in model robustness.

1.1 Adversarial Training for Object Detection

Adversarial training has been extensively explored in image classification, but its application to object detection introduces specific challenges due to the joint nature of classification and localization tasks. Lu et al. [12] were among the first to investigate adversarial attacks in object detection, showing that detection models are susceptible not only to classification errors but also to inaccuracies in bounding box predictions.

Subsequent approaches, such as those proposed by Xie et al. [25], adapted adversarial training from classification to detection by considering both classification loss and bounding box regression loss when crafting adversarial examples. However, these methods often face difficulties in balancing the trade-off between robustness in classification and precision in localization. Zhang and Wang [26] introduced a specialized method for object detection aimed at enhancing the robustness of both tasks.

An object detector $f(x) \rightarrow \{p_k, b_k\}_{k=1}^{K}$ processes an image $x \in [0, 255]^n$ and outputs up to K detected objects. Each object is represented by a probability vector $p_k \in \mathbb{R}^C$ for C classes (including the background) and a bounding box

$b_k = [x_k, y_k, w_k, h_k]$ [18]. The goal of training such a detector is to minimize the loss function over a training dataset D, given by:

$$\min_\theta \mathbb{E}_{(x, \{y_k, b_k\}) \sim D} \left[\text{loss}_{\text{cls}}(f_\theta(x), \{y_k\}) + \text{loss}_{\text{loc}}(f_\theta(x), \{b_k\}) \right] \tag{1}$$

where loss_{cls} is the classification loss and loss_{loc} is the localization loss.

Zhang and Wang proposed an adversarial training algorithm named MTD to improve the robustness of object detection model. Their approach decomposes the adversarial training process into two task-specific domains: S_{cls} for the classification task domain and S_{loc} for the regression task domain of the object detection loss as shown in equation (2) and equation (3).

$$S_{\text{cls}} = \{\bar{x} \mid \arg\max_{\bar{x} \in S_x} \text{loss}_{\text{cls}}(f(\bar{x}), \{y_k\})\} \tag{2}$$

$$S_{\text{loc}} = \{\bar{x} \mid \arg\max_{\bar{x} \in S_x} \text{loss}_{\text{loc}}(f(\bar{x}), \{b_k\})\} \tag{3}$$

where S_x denotes the ℓ_∞ ball around the clean image x with perturbation budget ϵ as shown in equation (4).

$$S_x = \{z \mid z \in B(x, \epsilon) \cap [0, 255]^n\}. \tag{4}$$

In adversarial training, as shown in equation (5), the objective is to address a complex minimax optimization problem, where the inner maximization aims to generate adversarial examples that exacerbate the model's loss, while the outer minimization seeks to optimize the model parameters to enhance overall robustness [7, 13, 22].

$$\min_\theta \max_{\|\delta\|_p \leq \epsilon} L(f_\theta(x + \delta), \{y_k, b_k\}) \tag{5}$$

where the perturbation δ is constrained within an ℓ_p-norm ball of radius ϵ. This constraint ensures that the adversarial perturbations are imperceptible while still maximizing the loss. Zhang and Wang presented an adversarial training approach that adheres to the task-specific domain constraints $S_{\text{cls}} \cup S_{\text{loc}}$. This approach generates adversarial examples for object classification and bounding box regression tasks and selects the example that maximizes the overall object detection loss, as shown in equation (6).

$$\min_\theta \left[\max_{\bar{x} \in S_{\text{cls}} \cup S_{\text{loc}}} L(f_\theta(\bar{x}), \{y_k, b_k\}) \right] \tag{6}$$

The adversarial training algorithm generates task-specific adversarial examples to enhance the model's robustness against classification and localization attacks. Specifically, as shown in equation (7), adversarial examples for the classification task ($\bar{x}_{\text{cls}}$) are generated by perturbing the input image in the direction of the gradient of the classification loss. This perturbation is scaled by the attack budget ϵ and projected back into the image space, ensuring the adversarial example remains within the valid input bounds. Similarly, as shown in equation (8), adversarial examples for the localization task ($\bar{x}_{\text{loc}}$) are created by focusing on the gradient of the localization loss, challenging the model's bounding box predictions.

$$\bar{x}_{\text{cls}} \overset{\triangle}{=} P_{S_x} \left(\tilde{x} + \epsilon \cdot \text{sign}\left(\nabla_x \text{loss}_{\text{cls}}(\tilde{x}, \{y_k\})\right)\right), \tag{7}$$

$$\bar{x}_{\text{loc}} \overset{\triangle}{=} P_{S_x} \left(\tilde{x} + \epsilon \cdot \text{sign}\left(\nabla_x \text{loss}_{\text{loc}}(\tilde{x}, \{b_k\})\right)\right), \tag{8}$$

The final adversarial example $\bar{x}$ is selected as the one that maximizes the overall detection loss across both tasks, ensuring that the training process balances robustness in classification and localization, as shown in equation (9). This approach prevents the model from being overly specialized in defending one task at the expense of the other, leading to improved robustness across the entire object detection pipeline.

$$\bar{x} = \arg \max_{\bar{x} \in \{\bar{x}_{\text{cls}}, \bar{x}_{\text{loc}}\}} L(f_\theta(\bar{x}), \{y_k, b_k\}). \tag{9}$$

By employing these formulas and methods, Zhang and Wang propose an effective adversarial training approach named MTD that enhances the robustness of object detection models against adversarial attacks.

1.2   Balanced Class Loss for Adversarial Training

Libra R-CNN, proposed by Pang et al. [3], introduced balanced learning for object detection. This method tackles the issue of class imbalance by integrating IoU-based losses and balanced feature pyramids, ensuring that training is equitably distributed among different classes. Such techniques are critical in maintaining adversarial robustness, as they prevent the model from being overly sensitive to specific classes.

Chen et al. [1] introduced Class-Aware Adversarial Training (CWAT), a novel method to improve the robustness of object detectors against adversarial attacks. While most adversarial defenses have traditionally focused on classification tasks, CWAT is specifically designed for object detection. It effectively balances class loss, preventing any single class from disproportionately influencing the overall detection loss.

The proposed method CWAT generates a universal adversarial perturbation that targets all objects in an image simultaneously by maximizing the loss for each object. Rather than aggregating losses across all objects, the total loss is decomposed into class-wise losses and normalized by the number of objects in each class. This ensures balanced robustness across all object classes, preventing any single class or object from disproportionately influencing the overall loss. The object detection loss is comprised of a classification loss and a regression loss for bounding box predictions. The total loss for all objects in an image is formulated as equation (10),

$$L_{C'} = \frac{1}{C} \sum_{c=1}^{C} \frac{1}{n_c} \sum_{j=1}^{n_c} \left(\hat{l}_{cls}^o(O_j, \{y_j\}, \theta) + \hat{l}_{reg}^o(O_j, \{b_j\}, \theta)\right) \tag{10}$$

where C is the number of classes in one image, and n_c is the number of matched default boxes in class c. N_o denotes the number of matched objects, l_{cls} represents the classification loss, l_{reg} denotes the regression loss, O_j refers to the matched objects (default boxes).

To ensure that no single object or class disproportionately influences the loss during adversarial training, the loss for each object is clipped. This mechanism prevents any task-specific loss from becoming excessively large formulated as equation (11),

$$l_{\text{task}}(x + \delta, \{y, b\}, \theta) = \min\left(l_{\text{task}}(x + \delta, \{y, b\}, \theta), \beta\right) \tag{11}$$

where β is a predefined clipping threshold. This clipping strategy helps maintain a balanced adversarial training process across all objects and tasks.

The adversarial training process is also framed as a min-max optimization problem, as shown in equation (5). The CWAT approach effectively enhances the robustness of object detectors by ensuring balanced adversarial defenses across various objects and classes, thereby improving performance in real-world scenarios.

2 Methodology

2.1 Improved Multi-task Adversarial Training for Object Detection

Confidence Loss of Object Detection. In traditional object detection tasks, the model's loss function is typically divided into two components: classification and localization. Each task corresponds to specific loss terms, commonly referred to as classification loss and regression loss. However, the loss can also be further broken down into classification loss, boxing loss, and confidence loss.

The confidence loss measures the likelihood that a predicted bounding box contains an object. This is typically calculated using the Binary Cross-Entropy (BCE) loss [17], which compares the predicted objectness score with the ground truth indicating whether an object is present or not. The formula for confidence loss can be written as equation (12),

$$\text{loss}_{\text{conf}} = \sum_i \text{BCE}(p_i^{\text{conf}}, t_i^{\text{conf}}) \cdot \text{balance}[i] \tag{12}$$

where p_i^{conf} represents the confidence score (objectness) of the i-th predicted bounding box, indicating the probability of an object being present in the box. t_i^{conf} is the target confidence score for the i-th bounding box, typically derived from the Intersection over Union (IoU) [6] with the ground truth box, reflecting the true presence of an object. $\text{BCE}(\cdot, \cdot)$ denotes the Binary Cross-Entropy loss function, which computes the discrepancy between the predicted and actual confidence values. $\text{balance}[i]$ is a balancing factor to weight the contribution of each prediction layer to the total loss. Below is the formula for generating adversarial samples using confidence loss:

$$\tilde{x}_{\text{conf}} = P_{S_x}\left(\tilde{x} + \epsilon \cdot \text{sign}\left(\nabla_x \text{loss}_{\text{conf}}(\tilde{x}, \{y_k\})\right)\right) \tag{13}$$

The confidence loss, together with the classification and boxing losses, forms the complete loss function in object detection, ensuring that the model not only accurately identifies the objects' categories but also correctly determines whether or not an object exists within the predicted bounding boxes.

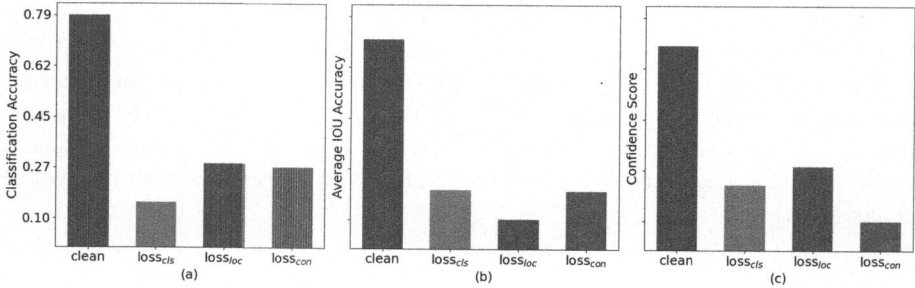

Fig. 1. The mutual impact of task losses, including classification accuracy, boxing average IOU accuracy, and confidence scores, is illustrated. (a) shows the classification accuracy of the model under various conditions: clean samples, adversarial samples generated based on classification loss (loss$_{cls}$), localization loss (loss$_{box}$), and confidence loss (loss$_{conf}$). (b) presents the boxing accuracy of the model for the same samples under different attack conditions. (c) displays the confidence scores of the model across various attacks.

Interactions of Different Task Losses. In object detection models, classification and localization tasks are inherently intertwined due to their shared base network and their combined influence during the non-maximum suppression (NMS) [19] phase. Our empirical study reveals that adversarial attacks tailored for one task often diminish the performance of the other task. For instance, an adversarial example generated using the classification loss ($loss_{cls}$) not only degrades classification accuracy but adversely affects boxing accuracy and confidence score. This mutual impact highlights that perturbations against one task propagate disturbances into other tasks, cultivating a cross-task attack transfer phenomenon (Fig. 1).

Not Fully Aligned Gradient Between Losses. Our analysis shows that the gradients of the classification and localization tasks have common directions but are not fully aligned, leading to conflicts that complicate adversarial training. To illustrate, we examined the image gradients derived from the classification, boxing and confidence losses ($g_c = \nabla_x loss_{cls}$, $g_l = \nabla_x loss_{box}$ and $g_c = \nabla_x loss_{con}$), uncovering that the magnitudes and directions of these task gradients are inconsistent. As shown in Figure 2, this misalignment results in loss gradients that intersect non-diagonally and exhibit distinct value ranges, demonstrating both overlap and distinctive regions in the gradient domains of the respective tasks.

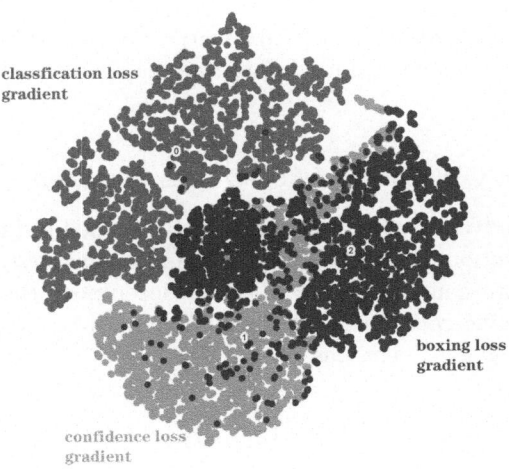

Fig. 2. Visualization of gradients for classification loss, boxing loss, and confidence loss using t-SNE. Under the PGD-20 attack budget, each point represents the gradient vector during the generation of adversarial examples. The colors indicate the loss types used for adversarial example generation: red for loss_{cls}, blue for loss_{loc}, and green for loss_{con}. Overlap among points reflects the interactions between different task losses. (Color figure online)

Our algorithm name MTDC decomposes the adversarial training into three task-oriented domains: S_{cls} for the classification branch, S_{box} for the boxing branch, and S_{conf} for the confidence branch of the object detection loss as shown in equation (14).

$$S_{\text{cls}} = \{\bar{x} \mid \arg\max_{\bar{x}\in S_x} \text{loss}_{\text{cls}}(f(\bar{x}), \{y_k\})\}$$

$$S_{\text{box}} = \{\bar{x} \mid \arg\max_{\bar{x}\in S_x} \text{loss}_{\text{box}}(f(\bar{x}), \{b_k\})\} \tag{14}$$

$$S_{\text{conf}} = \{\bar{x} \mid \arg\max_{\bar{x}\in S_x} \text{loss}_{\text{conf}}(f(\bar{x}), \{b_k\})\}$$

The adversarial training algorithm MTDC involves generating adversarial examples, as shown in equation (15), within each loss and selecting the one that maximizes the overall loss for training.

$$\bar{x}_{\text{cls}} \stackrel{\triangle}{=} P_{S_x}\left(\tilde{x} + \epsilon \cdot \text{sign}\left(\nabla_x \text{loss}_{\text{cls}}(\tilde{x}, \{y_k\})\right)\right)$$

$$\bar{x}_{\text{box}} \stackrel{\triangle}{=} P_{S_x}\left(\tilde{x} + \epsilon \cdot \text{sign}\left(\nabla_x \text{loss}_{\text{box}}(\tilde{x}, \{b_k\})\right)\right) \tag{15}$$

$$\bar{x}_{\text{conf}} \stackrel{\triangle}{=} P_{S_x}\left(\tilde{x} + \epsilon \cdot \text{sign}\left(\nabla_x \text{loss}_{\text{conf}}(\tilde{x}, \{b_k\})\right)\right)$$

By decomposing the overall loss into classification, bounding box, and confidence components, the original formulation in equation (9) is expanded into the more detailed expression in equation (16).

$$\bar{x} = \arg \max_{\bar{x} \in \{\bar{x}_{\mathrm{cls}}, \bar{x}_{\mathrm{box}}, \bar{x}_{\mathrm{conf}}\}} L(f_\theta(\bar{x}), \{y_k, b_k\}) \tag{16}$$

This extended formulation allows the adversarial training process to address each task-specific loss separately, improving the model's ability to resist adversarial attacks.

2.2   Integrating CWAT into MTDC Loss Function

CWAT is a novel adversarial training paradigm designed to improve the robustness of object detection models by addressing individual class-wise losses within an image. The core concept of CWAT is to decompose the overall detection loss into class-specific components, ensuring balanced and effective adversarial perturbations across all object classes.

$$L_t' = \frac{1}{C} \sum_{c=1}^{C} \frac{1}{n_c} \sum_{j=1}^{n_c} \left(\hat{l}_t^o(O_j, \{y_j\}, \theta) \right), \quad t \in \{\mathrm{cls}, \mathrm{box}, \mathrm{conf}\} \tag{17}$$

In Eq. (17), the parameter C denotes the total number of object classes in the image, while n_c represents the number of matched objects or bounding boxes for a given class c. The term O_j refers to the j-th object (or matched bounding box) in the image, and $\{y_j\}$ represents the set of ground truth labels for that object.

The CWAT method employs Equation (10) to compute the total loss for all objects in batch images. By decomposing the task-domain loss into three components, the loss for each type across all objects is calculated using Equation (17). Unlike approaches that normalize total loss across all objects, CWAT generates a universal adversarial perturbation that maximizes the loss for each class individually.

In MTDC-CWAT, CWAT is integrated into the MTDC loss function, where the loss_{cls}, loss_{box}, and loss_{conf} are computed using Equation (17) for all objects in a batch of images. For instance, when calculating the loss_{conf} individually, the total loss_{conf} is first summed over all objects, normalized by the number of objects, and finally averaged across the number of classes.

2.3   Fast Adversarial Training for Object Detection

While the proposed MTDC-CWAT enhances the robustness of object detection models, these techniques also introduce additional computational complexity. Specifically, MTDC requires calculating and optimizing multiple task-specific losses (classification, boxing, and confidence), which leads to an increase in the number of forward and backward passes during training. Each training iteration needs to generate adversarial examples and compute gradients for each task, resulting in higher computational overhead compared to standard training methods. Additionally, integrating CWAT further amplifies the overhead as it involves class-wise perturbations, ensuring that the robustness is evenly distributed across all object classes, which requires additional computation per class. Shafahi et al. [20] introduced Fast FGD, an accelerated variant of the PGD

Algorithm 1. Fast MTD-CWAT Adversarial Training with Class-Aware Balancing

---

Require: Training samples X, perturbation bound ϵ, learning rate τ, number of hop steps m

1: Initialize θ
2: $\delta \leftarrow 0$
3: **for** epoch $= 1$ to N_{ep}/m **do**
4:     **for** minibatch B from X **do**
5:         **for** i $= 1$ to m **do**
6:             Update θ with stochastic gradient descent
7:             $g_\theta \leftarrow \mathbb{E}_{(x,y)\in B}[\nabla_\theta l(x+\delta, y, \theta)]$
8:             Compute adversarial gradient: $g_{adv} \leftarrow \nabla_x l(x+\delta, y, \theta)$
9:             Update θ
10:           $\theta \leftarrow \theta - \tau \cdot g_\theta$
11:           Use gradients calculated for the minimization step to update δ
12:           $\delta \leftarrow \delta + \epsilon \cdot \text{sign}(g_{adv})$
13:           $\delta \leftarrow \text{clip}(\delta, -\epsilon, \epsilon)$
14:           Classification attack: $\bar{x}_i^{cls} = P_{S_x}(x+\delta+\epsilon \cdot \text{sign}(\nabla_x L'_{cls}(x+\delta, y)))$
15:           Boxing attack: $\bar{x}_i^{box} = P_{S_x}(x+\delta+\epsilon \cdot \text{sign}(\nabla_x L'_{box}(x+\delta, b)))$
16:           Confidence attack: $\bar{x}_i^{conf} = P_{S_x}(x+\delta+\epsilon \cdot \text{sign}(\nabla_x L'_{conf}(x+\delta, b)))$
17:           $L_{cls} = L(\bar{x}_i^{cls}, \{y, b\})$
18:           $L_{loc} = L(\bar{x}_i^{loc}, \{y, b\})$
19:           $L_{con} = L(\bar{x}_i^{con}, \{y, b\})$
20:           $\bar{x}_i = \bar{x}_j$ where $j = \arg\max_{j \in \{cls, box, conf\}}\{L_{cls}, L_{box}, L_{conf}\}$
21:           $\delta \leftarrow \delta + \epsilon \cdot \text{sign}(\nabla_x L(\bar{x}_i, y, \theta))$
22:           $\delta \leftarrow \text{clip}(\delta, -\epsilon, \epsilon)$
23:         **end for**
24:     **end for**
25: **end for**
26: **Output:** The learned model parameters θ

---

algorithm for adversarial training. FGD improves the generation of adversarial examples by reusing gradient information from the model's parameter updates, thus eliminating the need for multiple gradient computations. This simultaneous backward pass reduces computational cost without compromising robustness, making FGD up to 30 times faster than traditional PGD while maintaining comparable performance. By incorporating fast adversarial training techniques, we mitigate some of the computational cost, achieving a balance between robustness and training efficiency.

Finally, the algorithm integrates MTD, CWAT, and FGD to enhance its performance. The specifics of the MTDC-CWAT adversarial training method for object detection are detailed in Algorithm 1.

3 Experiment

3.1 Datasets and Evaluation Settings

In our experiments, we adopt the widely-established "07+12" protocol for the PASCAL VOC dataset, utilizing a combined set of 16,551 images from the VOC 2007 and 2012 datasets, which include 40,058 annotated objects spanning 20 distinct classes for training purposes. For evaluation, we employ the PASCAL VOC 2007 test set, which consists of 4,952 images.

For the MS-COCO dataset, we use the 2017 training set, comprising 118,287 images across 80 object categories. The evaluation is performed on the 2017 validation set, containing 5,000 images. To evaluate the robustness of the detector, we utilize the mAP metric with an Intersection over Union (IoU) [4,6,11] threshold of 0.5, providing insight into the model's performance under varying conditions.

3.2 Implementation Details

For our experiments, we implement a comprehensive parameter tuning strategy to enhance the performance and robustness of object detection models. We choose YOLOv5s as the base object detection framework due to its highly efficient architecture, which balances speed and accuracy.

The YOLOv5s model is used as the primary framework, with an initial learning rate set to 10^{-2} and decayed by a factor of 0.1 at specified iterations to promote effective convergence. We utilize a batch size of 32 to balance computational efficiency with gradient stability. The adversarial attack budget is set to $\epsilon = 8$, ensuring a practical balance between adversarial robustness and visual similarity to the original images. The input image size for YOLOv5s is configured to 640×640 pixels to match the model's architectural requirements. The pixel values are initially within the range of [0, 255] and are then normalized by subtracting the mean pixel intensity computed over the entire dataset.

The SGD optimizer is employed with a momentum of 0.6 and a weight decay of 0.001 to facilitate effective learning and mitigate overfitting. The learning rate is adjusted at 40k, 60k, and 80k iterations for the PASCAL VOC dataset, and at 180k, 220k, and 260k iterations for the MS-COCO dataset. For our Fast-PGD-based adversarial training, we set $m = 4$.

The clean model is trained for 100 epochs, in the adversarial training setup, the first 50 epochs utilize only clean samples, while the remaining 50 epochs employ a mixed approach where each batch contains 50% clean samples and 50% adversarial samples.

3.3   Evaluation Results on Pascal VOC and MS-COCO

In this subsection, we present the evaluation results on both the Pascal VOC and MS-COCO datasets. The MS-COCO dataset, being more representative of real-world object detection scenarios, poses greater challenges in evaluating the robustness of object detectors. Consequently, it serves as a more rigorous benchmark compared to Pascal VOC, which contains fewer objects per image and fewer object classes, making it less complex. By comparing the performance on both datasets, we can better assess the effectiveness of various adversarial training methods in diverse detection environments.

We evaluate the performance of five different models under FGSM and PGD-20 attacks. The specific models and their details are outlined belows:

- **STD**: The object detector trained with natural training using clean images.
- **MTD**: The model trained following the algorithm described in [26].
- **CWAT**: The model trained with samples generated using CWAT loss during adversarial training [1].
- **CMTD**: The model trained using improved MTD.
- **CMTD-CWAT**: The model trained using Algorithm 1 (Tables 1 and 2).

Table 1. The evaluation results of various adversarially trained YOLOv5 models under FGSM, PGD-20 attacks with $\epsilon = 8$ on the PASCAL VOC test set.

attacks	clean	FGSM			PGD-20		
		$loss_{cls}$	$loss_{box}$	$loss_{conf}$	$loss_{cls}$	$loss_{box}$	$loss_{conf}$
STD	0.784	0.0275	0.0246	0.030	0.002	0.004	0.004
MTD	0.636	0.457	0.456	0.453	0.400	0.404	0.396
CWAT	0.686	0.416	0.415	0.357	0.362	0.353	0.301
MTD3	0.622	0.461	0.461	0.458	0.410	0.409	0.406
MTD3-CWAT	0.642	**0.479**	**0.478**	**0.463**	**0.427**	**0.429**	**0.417**

Table 2. The evaluation results of various adversarially trained YOLOv5 models under FGSM, PGD-20 attacks with $q = 8$ on the MS-COCO test set.

attacks	clean	FGSM			PGD-20		
		$loss_{cls}$	$loss_{box}$	$loss_{conf}$	$loss_{cls}$	$loss_{box}$	$loss_{conf}$
STD	0.504	0.022	0.020	0.025	0.002	0.002	0.002
MTD	0.499	0.203	0.213	0.160	0.201	0.202	0.152
CWAT	0.487	0.224	0.235	0.191	0.229	0.231	0.179
MTD3	0.459	0.285	0.278	0.277	0.272	0.272	0.271
MTD3-CWAT	0.478	**0.322**	**0.318**	**0.319**	**0.302**	**0.301**	**0.298**

On the PASCAL VOC test set, the MTD3 model exhibits superior performance over the MTD model under adversarial attacks. While MTD3 shows a marginally lower mAP on clean samples, with a difference of 0.014 compared to MTD, its robustness under adversarial attacks is significantly enhanced. Notably, under the PGD-20 attack, MTD3 improves the mAP by 0.010, 0.005, and 0.010, respectively. Moreover, the MTD3-CWAT model achieves the highest mAP under both FGSM and PGD-20 attacks, demonstrating stronger robustness across various adversarial scenarios. CWAT performs better than MTD on the MS-COCO dataset, though it underperforms compared to MTD on the PASCAL VOC dataset. This difference arises because the MS-COCO dataset more accurately mirrors real-world object detection scenarios, offering a more challenging benchmark for evaluating the robustness of object detectors [1]. On the MS-COCO dataset, MTD3-CWAT also surpasses both MTD and CWAT under PGD-20 adversarial attacks.

3.4 Ablation Study

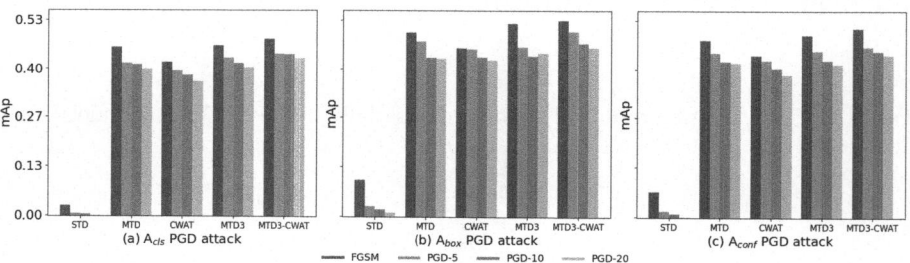

Fig. 3. Robustness performance of each model under adversarial attacks with $\epsilon = 8$ across varying numbers of PGD steps on the PASCAL VOC 2007 test set. The STD model represents the standard baseline, while MTD, CWAT, MTD3, and MTD-CWAT denote the robust models proposed.

Attacks Under Different Number of PGD Steps. We start by assessing the performance of the models under adversarial attacks with varying numbers of PGD steps in PASCAL VOC 2007 test set, while keeping the attack budget with $\epsilon = 8$. The results, shown in Fig. 3, reveal several key insights: *i)* The performance of the STD model deteriorates rapidly after only a few PGD steps and continues to decline sharply, approaching zero as the number of steps increases. This trend is consistent across attacks based on $loss_{cls}$, $loss_{box}$, and $loss_{conf}$, highlighting the effectiveness of these adversarial attack techniques in undermining detector performance. *ii)* In contrast, the robust models demonstrate significantly more stable performance across varying numbers of PGD steps, reflecting their enhanced resilience against adversarial attacks compared to the standard model. These findings underscore the vulnerability of the standard models and the superior robustness of the advanced models in maintaining performance under adversarial conditions.

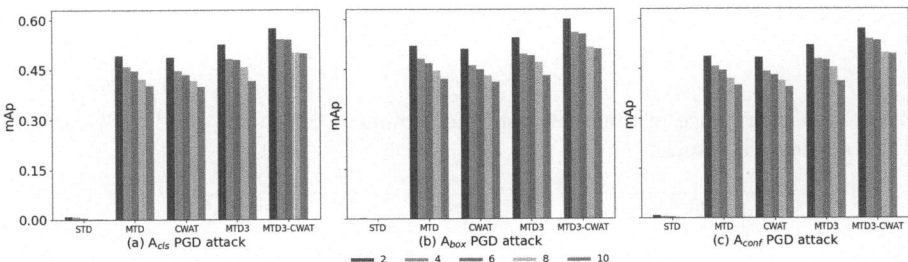

Fig. 4. Robustness of each model under varying attack budgets, with the number of PGD steps fixed at 20, on the PASCAL VOC 2007 test set. STD represents the standard model, while MTD, CWAT, MTD3, and MTD-CWAT denote our robust models.

Attacks Under Different Budgets. We evaluate the robustness of models under varying attack budgets, specifically for $\epsilon \in \{2, 4, 6, 8, 10\}$, with the number of PGD steps fixed at 20, as illustrated in Fig. 4. The results indicate a significant deterioration in the performance of STD, trained on natural images, as the attack budget increases. Notably, in A_{cls} PGD attack, its accuracy decreases markedly from 0.695 on clean images to around 0.03 with an attack budget of $\epsilon = 2$. The performance of the STD model remains poor even at $\epsilon = 2$. In contrast, robust models exhibit a more gradual decline in performance as the attack budget increases, demonstrating substantially better resilience to adversarial perturbations. Among the robust models, MTD3-CWAT achieves the best performance, followed by MTD3, regardless of the attack budget. These findings underscore the considerable difference in robustness between models trained on natural images and those optimized for adversarial defense.

Evaluation of Different Loss Combinations. We conducted ablation experiments to evaluate the effectiveness of various combinations of loss functions in generating adversarial samples for multi-task domain adversarial training. Specifically, we assessed six distinct combinations of the following loss functions: $loss_{cls}$, $loss_{box}$, and $loss_{conf}$. The combinations tested were: $loss_{cls}$, $loss_{box}$, $loss_{conf}$, $loss_{cls} + loss_{box}$, $loss_{cls} + loss_{conf}$, $loss_{box} + loss_{conf}$, and $loss_{cls} + loss_{box} + loss_{conf}$. The models and their corresponding loss functions are described as follows:

- **CLS**: uses S_{cls} exclusively for training.
- **BOX**: uses S_{box} exclusively for training.

- **OBJ**: uses S_{conf} exclusively for training.
- **BOX-CLS**: uses $S_{\text{cls}} \cup S_{\text{box}}$ for training.
- **BOX-OBJ**: uses $S_{\text{box}} \cup S_{\text{conf}}$ for training.
- **OBJ-CLS**: uses $S_{\text{conf}} \cup S_{\text{cls}}$ for training.
- **CON**: uses $S_{\text{cls}} \cup S_{\text{box}} \cup S_{\text{conf}}$ for training.

Table 3. Performance of different task loss combinations on the MS-COCO test set under adversarial attacks ($\epsilon = 8$)

attacks	clean	FGSM			PGD-20		
		$loss_{\text{cls}}$	$loss_{\text{box}}$	$loss_{\text{obj}}$	$loss_{\text{cls}}$	$loss_{\text{box}}$	$loss_{\text{obj}}$
CON	0.516	0.181	0.183	0.136	0.174	0.75	0.132
CLS	0.503	0.241	0.233	0.197	0.232	0.221	0.192
BOX	0.488	0.191	0.203	0.149	0.185	0.196	0.147
OBJ	0.508	**0.245**	**0.252**	**0.218**	**0.236**	**0.240**	**0.209**
BOX-CLS	0.500	0.200	0.210	0.156	0.191	0.190	0.149
BOX-OBJ	0.494	0.199	0.210	0.157	0.188	0.198	0.153
OBJ-CLS	0.509	**0.251**	**0.257**	**0.218**	**0.243**	**0.242**	**0.207**
MTD	0.499	0.203	0.213	0.160	0.201	0.202	0.152
CWAT	0.487	0.224	0.235	0.191	0.229	0.231	0.179
MTD3	0.459	0.285	0.278	0.277	0.272	0.272	0.271
MTD3-CWAT	0.478	**0.322**	**0.318**	**0.319**	**0.302**	**0.301**	**0.298**

The results, summarized in Table 3, indicate that different task domains lead to varying levels of model robustness. Specifically, among methods employing a single task domain, CLS demonstrates the highest robustness, followed by OBJ, while BOX exhibits the lowest robustness among single-task domains. For combined task domains, OBJ-CLS performs best, followed by BOX-CLS. These findings highlight the essential role of confidence loss in improving the model's adversarial robustness, both when used independently and in combination with other loss functions.

4   Conclusion

In this study, we propose a comprehensive approach to enhancing the robustness of object detection models against adversarial attacks. Through a systematic analysis of vulnerabilities in object detectors, we introduce several innovative techniques, including a confidence loss mechanism, class-wise adversarial training, and an accelerated adversarial training method. Our findings demonstrate the effectiveness of categorizing adversarial training loss into three distinct components and highlight the critical role of confidence loss in improving model robustness. Extensive evaluations on the PASCAL VOC and MS-COCO datasets reveal significant improvements in model resilience and performance consistency across various object classes.

While this work marks a significant advancement, future research should address challenges such as developing advanced loss functions, optimizing the trade-offs between clean image accuracy and robustness, and refining multi-task learning approaches. Furthermore, our framework can be extended to enhance the robustness of other multi-task models, establishing a solid foundation for further exploration into adversarial robustness in object detection and beyond.

Acknowledgments. This work was supported by High-level Talent Foundation of Hebei Province (Grant No. C20221062), Military-Civilian Tntegration Development Foundation of Hebei Province (Grant No. HB24JMRH026) and Doctor Scientific Research Foundation of Shijiazhuang University (Grant No.18BS013).

References

1. Chen, P.C., Kung, B.H., Chen, J.C.: Class-aware robust adversarial training for object detection. In: Proceedings of the IEEE/CVF Conference on Computer Vision and Pattern Recognition (CVPR), pp. 10420–10429 (2021)
2. Chen, S.T., Cornelius, C., Martin, J., Chau, D.H.: ShapeShifter: robust physical adversarial attack on faster R-CNN object detector. In: Machine Learning and Knowledge Discovery in Databases: European Conference, ECML PKDD 2018, Dublin, Ireland, September 10–14, 2018, Proceedings, Part I 18, pp. 52–68. Springer (2019)
3. Cheng, Q., Wu, Y., Chen, F., Guo, Y.: Balanced loss for accurate object detection. In: Peng, Y., et al. (eds.) Pattern Recognition and Computer Vision, pp. 342–354. Springer, Cham (2020)
4. Everingham, M., Eslami, S.A., Van Gool, L., Williams, C.K., Winn, J., Zisserman, A.: The Pascal visual object classes challenge: a retrospective. Int. J. Comput. Vis. **111**, 98–136 (2015)
5. Fawzi, A., Fawzi, O., Frossard, P.: Analysis of classifiers' robustness to adversarial perturbations. Mach. Learn. **107**(3), 481–508 (2018)
6. Girshick, R., Donahue, J., Darrell, T., Malik, J.: Rich feature hierarchies for accurate object detection and semantic segmentation. In: Proceedings of the IEEE Conference on Computer Vision and Pattern Recognition, pp. 580–587 (2014)
7. Goodfellow, I.J., Shlens, J., Szegedy, C.: Explaining and harnessing adversarial examples. arXiv preprint arXiv:1412.6572 (2014)
8. Guo, F., et al.: Detecting adversarial examples via prediction difference for deep neural networks. Info. Sci. **501**, 182–192 (2019). https://doi.org/10.1016/j.ins.2019.05.084
9. Ilyas, M., Khaw, H.Y., Selvaraj, N.M., Jin, Y., Zhao, X., Cheah, C.C.: Robot-assisted object detection for construction automation: data and information-driven approach. IEEE/ASME Trans. Mechatron. **26**(6), 2845–2856 (2021)
10. Jiao, L., et al.: A survey of deep learning-based object detection. IEEE Access **7**, 128837–128868 (2019)
11. Lin, T.Y., et al.: Microsoft CoCo: common objects in context. In: Computer Vision–ECCV 2014: 13th European Conference, Zurich, Switzerland, September 6-12, 2014, Proceedings, Part V 13, pp. 740–755. Springer (2014)

12. Lu, J., Sibai, H., Fabry, E., Forsyth, D.: No need to worry about adversarial examples in object detection in autonomous vehicles. arXiv preprint arXiv:1707.03501 (2017)
13. Madry, A.: Towards deep learning models resistant to adversarial attacks. arXiv preprint arXiv:1706.06083 (2017)
14. Mi, J.X., Wang, X.D., Zhou, L.F., Cheng, K.: Adversarial examples based on object detection tasks: a survey. Neurocomputing **519**, 114–126 (2023)
15. Moosavi-Dezfooli, S.M., Fawzi, A., Frossard, P.: DeepFool: a simple and accurate method to fool deep neural networks. In: Proceedings of the IEEE Conference on Computer Vision and Pattern Recognition, pp. 2574–2582 (2016)
16. Pang, T., Yang, X., Dong, Y., Xu, K., Zhu, J., Su, H.: Boosting adversarial training with hypersphere embedding. Adv. Neural. Inf. Process. Syst. **33**, 7779–7792 (2020)
17. Redmon, J., Divvala, S., Girshick, R., Farhadi, A.: You only look once: unified, real-time object detection. In: Proceedings of the IEEE Conference on Computer Vision and Pattern Recognition, pp. 779–788 (2016)
18. Ren, S.: Faster R-CNN: towards real-time object detection with region proposal networks. arXiv preprint arXiv:1506.01497 (2015)
19. Rosenfeld, A., Thurston, M.: Edge and curve detection for visual scene analysis. IEEE Trans. Comput. **100**(5), 562–569 (1971)
20. Shafahi, A., et al.: Adversarial training for free! Adv. Neural Info. Process. Syst. **32** (2019)
21. Song, Z., et al.: Robustness-aware 3D object detection in autonomous driving: a review and outlook. IEEE Trans. Intell. Transp. Syst. (2024)
22. Szegedy, C.: Intriguing properties of neural networks. arXiv preprint arXiv:1312.6199 (2013)
23. Wang, L., et al.: Multi-modal 3D object detection in autonomous driving: a survey and taxonomy. IEEE Trans. Intell. Veh. **8**(7), 3781–3798 (2023)
24. Wang, Y., Tan, Y.A., Zhang, W., Zhao, Y., Kuang, X.: An adversarial attack on DNN-based black-box object detectors. J. Netw. Comput. Appl. **161**, 102634 (2020). https://doi.org/10.1016/j.jnca.2020.102634
25. Xie, C., Wang, J., Zhang, Z., Zhou, Y., Xie, L., Yuille, A.: Adversarial examples for semantic segmentation and object detection. In: Proceedings of the IEEE International Conference on Computer Vision, pp. 1369–1378 (2017)
26. Zhang, H., Wang, J.: Towards adversarially robust object detection. In: Proceedings of the IEEE/CVF International Conference on Computer Vision (ICCV) (2019)
27. Zhang, H., Li, D., Ji, Y., Zhou, H., Wu, W., Liu, K.: Toward new retail: a benchmark dataset for smart unmanned vending machines. IEEE Trans. Industr. Inf. **16**(12), 7722–7731 (2019)

FedHKD: A Hierarchical Federated Learning Approach Integrating Clustering and Knowledge Distillation for Non-IID Data

Shiwen Hu[1], Changji Wang[1,2(✉)], Yuan Li[1], Zhen Liu[1,2], Ning Liu[1,2], and Qingqing Gan[1,2]

[1] School of Information Science and Technology, Guangdong University of Foreign Studies, Guangzhou, China
wchangji@126.com
[2] Guangdong Engineering Research Center of Data Security Governance and Privacy Computing, Guangzhou, China

Abstract. Federated learning allows decentralized model training while preserving data privacy. However, Non-IID data poses significant challenges, leading to performance degradation and increased communication overhead. This paper introduces FedHKD, a hierarchical FL algorithm that integrates client clustering and knowledge distillation to address Non-IID challenges. By clustering clients with similar data distributions and transferring knowledge across clusters, FedHKD improves model accuracy while reducing communication costs. Experimental results on Fashion-MNIST and CIFAR-10 datasets demonstrate up to 14.24% improvement in accuracy and a substantial reduction in communication overhead, proving its effectiveness in Non-IID environments.

Keywords: Federated learning · Non-IID · knowledge distillation · clustering

1 Introduction

The rapid expansion of big data raises significant privacy concerns, particularly in scenarios where data sharing is restricted by regulations or proprietary limitations. Federated learning (FL) offers a decentralized approach that enables collaborative model training without the need to share raw data. Instead, FL operates by exchanging only model parameters between clients and a central server, thereby preserving data privacy and addressing data silo issues [1].

However, one of the key challenges in FL is handling Non-IID (non-independent and identically distributed) data, where the data distribution across clients varies significantly. This heterogeneity can lead to issues such as model divergence, reduced global model performance, and increased communication overhead between clients and the server [2]. Although various methods such as

F. Zhang et al. (Eds.): AIS&P 2024, LNCS 15399, pp. 105–116, 2025.
https://doi.org/10.1007/978-981-96-1148-5_9

FedProx [3] and SCAFFOLD [4] have been proposed to mitigate these challenges, they often face difficulties in highly Non-IID environments, and their performance is compromised by additional communication costs.

In this paper, we propose FedHKD, a novel hierarchical federated learning algorithm that integrates client clustering with knowledge distillation to effectively address the challenges posed by Non-IID data. By clustering clients with similar data distributions, FedHKD reduces the negative impact of data heterogeneity. Additionally, knowledge distillation is employed to facilitate efficient knowledge transfer across clusters, improving global model performance. By designating a cluster leader to manage communication with the server, FedHKD also significantly reduces communication overhead, enhancing scalability and efficiency.

The key contributions of this paper are as follows:

- We propose a novel hierarchical federated learning algorithm, FedHKD, which integrates client clustering and knowledge distillation to address Non-IID data challenges in federated learning.
- We demonstrate that FedHKD achieves substantial improvements in model performance, with accuracy gains of up to 14.24% on benchmark datasets such as Fashion-MNIST and CIFAR-10.
- FedHKD reduces communication overhead by leveraging cluster leaders, making the algorithm highly scalable and more efficient for large-scale federated learning deployments.

The rest of this paper is organized as follows: In Sect. 2, we review the related work on federated learning, particularly focusing on strategies for handling Non-IID data and reducing communication overhead. Section 3 presents the preliminaries, introducing key concepts in federated learning and knowledge distillation. In Sect. 4, we describe the proposed FedHKD algorithm in detail, including the client clustering process and the implementation of cross-cluster knowledge distillation. Section 5 discusses the experimental setup and provides a comprehensive evaluation of FedHKD, comparing its performance with existing federated learning algorithms. Finally, Sect. 6 concludes the paper and outlines potential directions for future work.

2   Related Work

Several aggregation algorithms have been proposed to address the challenges posed by Non-IID data in federated learning. For example, FedProx [3] introduces a regularization term to constrain local model updates, ensuring they remain closer to the global model, which helps mitigate the negative effects of data heterogeneity. Similarly, SCAFFOLD [4] incorporates control variates to correct the direction of client model updates, thereby improving convergence rates in Non-IID environments. Another approach, FedNova [5], normalizes and weights each client's local updates to reduce training bias caused by data heterogeneity. Zhu et al. [6] analyzed the pros and cons of these methods, concluding that while such aggregation algorithms can improve convergence speed and model stability, their effectiveness is still limited in scenarios where the data distributions across clients are highly diverse. Furthermore, these algorithms often increase communication overhead due to the frequent exchange of model updates between all clients and the server.

Beyond aggregation algorithms, additional techniques have been developed to address Non-IID data challenges. Yang et al. proposed the G-FML algorithm [7], which employs a meta-learning approach to extract meta-features from client models within a cluster and dynamically adjusts the models to enhance overall global model performance. Similarly, Ma et al. introduced the pFedLA algorithm [8], which assigns a specialized super network to each client, allowing the identification and prioritization of important model layers. This hierarchical aggregation method improves model performance but at the cost of increased communication overhead due to frequent interactions between clients and the server.

Moreover, innovative techniques like knowledge distillation and generative adversarial networks (GANs) have been integrated into federated learning to further alleviate Non-IID issues. Knowledge distillation facilitates model consistency by transferring knowledge from a larger, more complex teacher model to a smaller student model, enhancing overall performance. Lin et al. [9] proposed a knowledge distillation method based on adversarial training, improving model accuracy in Non-IID settings by enabling efficient knowledge transfer between the teacher model and both the clients and server. Wang et al. [10] introduced a dual knowledge distillation framework that further enhances model accuracy in Non-IID environments by implementing knowledge transfer between client and server models. Despite the potential improvements in model performance, these approaches also introduce challenges, as frequent communication between clients and servers is required to facilitate distillation, increasing communication overhead. Moreover, careful selection of teacher models and distillation strategies is necessary to optimize performance gains without overburdening the system.

3 Preliminaries

3.1 Federated Learning and FedAvg Algorithm

Federated learning is a distributed machine learning framework designed to protect data privacy. It enables multiple data sources to collaboratively train models without sharing the original data, effectively safeguarding privacy and eliminating data silos. In Federated learning, data remains stored locally, and only model parameters are exchanged between nodes. This approach not only enhances data privacy but also improves the model's generalization ability by leveraging multi-source data.

Handling Non-IID data is a fundamental challenge in federated learning. In real-world scenarios, there are often substantial variations in the data distribution among different clients, which can result in considerable discrepancies between local model updates. These variations may prompt the following questions:

– **Model drift:**When local models trained on Non-IID data are aggregated, the resulting global model may exhibit bias or instability, which can diminish its overall effectiveness.
– **Slowed convergence:** Inconsistencies in client updates on Non-IID data hinder the convergence of the global model, necessitating additional communication rounds to attain performance levels comparable to those in the IID setting.
– **Unbalanced performance:** The global model learns client-specific data features rather than common features across all data.

3.2   Knowledge Distillation

Knowledge distillation is a widely used technique in machine learning that aims to compress and optimize models by transferring knowledge from a complex model (the teacher model) to a simpler model (the student model). This process ensures that the student model performs comparably to the teacher model. In federated learning, knowledge distillation can be applied in the following ways:

– **Cross-cluster knowledge distillation:** By transferring knowledge between the teacher model and the student model across different clusters, the performance of the global model is significantly enhanced.
– **Local private model distillation:** Performing knowledge distillation locally on the client can ensure data privacy while also enhancing model performance.

4   Proposed Method

We propose the FedHKD algorithm, which addresses Non-IID data issues by integrating clustering and knowledge distillation. The process begins by clustering clients based on data distribution statistics (mean, standard deviation, skewness). The server uses evaluation metrics such as the Silhouette coefficient and Calinski-Harabasz index to determine the optimal number of clusters. Clients within each cluster communicate with the server through a designated cluster leader, reducing communication overhead. The FedHKD process is detailed in Algorithm 1.

4.1   Clustering Based on Client Similarity

Before training, the client must share the distribution statistics of its local data with the server. This step is crucial for addressing the challenges posed by Non-iid data distribution. Typically, the client will provide the mean μ_i, standard deviation σ_i, and skewness γ_i of its dataset to the server, in the following form:

$$clients\_statis = \{(\mu_1, \sigma_1, \gamma_1), \ldots, (\mu_n, \sigma_n, \gamma_n)\} \tag{1}$$

After receiving data distribution statistics from all clients, the server will utilize various evaluation metrics during the clustering process to determine the optimal number of clusters, K. These metrics include the Silhouette coefficient [11], the Calinski-Harabasz index [12], and the Davies-Bouldin index [13]. Each of these indicators assesses clustering quality across different numbers of clusters, facilitating the selection of an appropriate K value. By integrating the evaluation results from these three metrics, the server can identify the optimal number of clusters, K, to achieve effective client clustering while ensuring homogeneity within clusters and heterogeneity between them.

Algorithm 1. FedHKD: A Hierarchical Federated Learning Approach Integrating Clustering and Knowledge Distillation for Non-IID Data

---

Input: Client datasets $\{D_i\}_{i=1}^{n}$, number of clusters K, number of communication rounds T, public data set D_{pds}

Output: Next round global model w_g^{t+1}

Initialize phase:

Client i computes and sends statistics μ_i, σ_i, γ_i to the server

The server computes the optimal number of clusters K using the Silhouette coefficient, Calinski-Harabasz Index, and Davies-Bouldin Index

The server uses the KMeans algorithm to categorize clients into K clusters $\{C_k\}_{k=1}^{K}$ based on their statistical data

The server sends set information $\{C_k\}_{k=1}^{K}$ to all clients

Train phase:

for $t = 1$ to T **do**

    for each cluster $k \in \{1, ..., K\}$ **do**

        Clients in cluster C_k negotiate and select a cluster leader $C_k^{l(i)}$ based on computing and communication capabilities

        $C_k^{l(i)}$ sends global model w_g^t to clients in cluster C_k

        Clients update local models $w_t^{C_k^i}$ using local datasets D_k^i

        Clients send updated local models to $C_k^{l(i)}$

        $C_k^{l(i)}$ uses **FedAvg** to aggregate all client models in cluster C_k to form a cluster aggregation model $w_t^{C_k}$

        Cluster leader $C_k^{l(i)}$ sends cluster aggregation model $w_t^{C_k}$ to the server

    end for

    Server collects all cluster aggregation models $\{w_t^{C_k}, k = 1, \ldots, K\}$

    Server randomly selects a cluster aggregation model as teacher model $T_t^{C_k}$ and the rest as student models $S_t^{C_k}$

    Train student model $S_t^{C_k}$ using knowledge distillation with teacher model $T_t^{C_k}$

    Server uses **FedAvg** to aggregate all $S_t^{C_k}$ and $T_t^{C_k}$ to obtain global model w_g^{t+1}

end for

The server employs the KMeans algorithm to cluster the data distribution statistics of all clients. The essence of the KMeans algorithm lies in its iterative optimization of the cluster centers, ensuring that the samples within each cluster are as close to the cluster center as possible, thereby minimizing the squared error within the cluster. The objective is to partition the dataset into K clusters, with each cluster represented by its centroid. The mathematical formulation of the objective function is as follows:

$$\arg\min_{C} \sum_{i=1}^{K} \sum_{x \in C_i} \|x - \mu_i\|^2 \tag{2}$$

$C = \{C_1, C_2, \ldots, C_K\}$ represents a set of K clusters, x is a sample point in the data set, and μ_i is the centroid of cluster C_i (the mean vector of all points in the cluster).

4.2 Knowledge Sharing Among Clusters

In the FedHKD algorithm, the server randomly selects one cluster aggregation model to serve as the teacher model, while the remaining models function as student models. The specific steps are as follows:

(1) Teacher model selection: The server randomly selects one from the cluster aggregation model as the teacher model:

$$T_t^{C_k} \leftarrow \text{Random}\{w_t^{C_k}, k = 1, \ldots, K\} \tag{3}$$

(2) Teacher model knowledge sharing: In each round of training, the server shares the knowledge of the teacher model with the student model by calculating the distillation loss. Here, $\nabla\ell(\cdot)$ denotes the gradient of the loss function with respect to the model parameters.

$$S_t^{C_k} \leftarrow \arg\min \sum \nabla\ell\left(S_t^{C_k}, D_{pds}\right) + \text{DistillationLoss}\left(T_t^{C_k}(D_{pds}), S_t^{C_k}(D_{pds})\right) \tag{4}$$

Through knowledge distillation, the student model can acquire additional knowledge from the teacher model, thereby enhancing the performance of the global model.

5   Experimental Results

5.1   Experimental Setup

Environment and Configuration: The experiments were conducted on a high-performance server equipped with an Intel Xeon E5-2650 v4 CPU, NVIDIA Tesla P100 GPU, and 128 GB of memory, running Ubuntu 18.04 and TensorFlow 2.4.

Parameter Settings: We simulated a FL environment with 20 clients. The learning rate η was set to 0.01, and the batch size was 64. The optimal number of clusters K, was set to 4.

Datasets: We evaluated the proposed FedDAM algorithm using the Fashion-MNIST and CIFAR-10 datasets. Fashion-MNIST consists of 60,000 grayscale images of size 28x28, with 50,000 images for training and 10,000 for testing across 10 categories of fashion products. CIFAR-10 comprises 60,000 color images of size 32x32, also divided into 10 categories, with 50,000 images for training and 10,000 for testing.

Data Distribution Settings: To simulate a Non-IID environment, we used a Dirichlet distribution to generate the data distributions among clients. Specifically, given a dataset D with l categories and n clients, a probability vector p of length l was generated for each client, where $p \sim \text{Dir}(\alpha)$. Here, α controls the degree of Non-IID: a smaller α leads to higher degrees of Non-IID data among clients.

Model Architecture: For the Fashion-MNIST dataset, we utilized a convolutional neural network (CNN) consisting of four convolutional layers, followed by a flattening layer and a fully connected layer for classification. In contrast, for the CIFAR-10 dataset, we implemented a more complex CNN architecture that included two convolutional layers, two pooling layers, two fully connected layers, a flattening layer, and a dropout layer to improve generalization.

Baseline Setting: To demonstrate the superior performance of the FedHKD algorithm, this paper conducts a comprehensive comparison with three widely used federated learning algorithms: FedAvg, FedProx, and SCAFFOLD. Both FedProx and SCAFFOLD are robust aggregation algorithms specifically designed for processing Non-IID data. Additionally, we evaluate the performance of the FedHKD algorithm without knowledge distillation.

5.2   Experimental Results and Analysis

Test Accuracy: Figures 1 and 2 present the experimental results of the four algorithms on the Fashion-MNIST and CIFAR-10 datasets.

Results show that FedHKD significantly outperforms FedAvg in handling Non-IID data, addressing uneven distribution effectively, where FedAvg struggles. Although the FedProx algorithm shows some improvement over FedAvg in Non-IID conditions, this enhancement is limited. In contrast, the SCAFFOLD algorithm further optimizes adaptability to Non-IID data, outperforming both FedProx and FedAvg overall. However, under highly Non-IID conditions ($\alpha = 0.05$ and $\alpha = 0.1$), the performance of FedProx and SCAFFOLD still falls short of that of FedHKD, underscoring the significant advantage of FedHKD in processing highly Non-IID data.

Test Loss Value: Figures 3 and 4 illustrate the changes in loss values for four algorithms applied to the Fashion-MNIST and CIFAR-10 datasets, respectively,

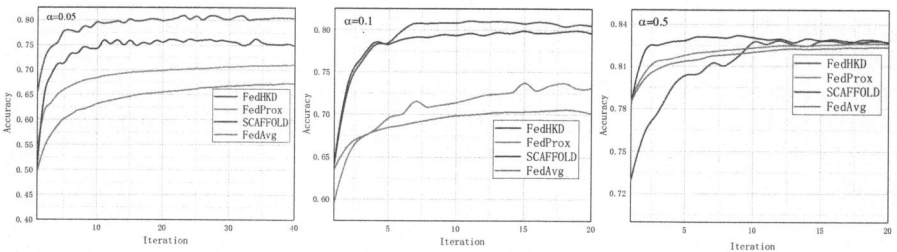

Fig. 1. Test Accuracy on the Fashion-MNIST Dataset under Varying Non-IID Levels

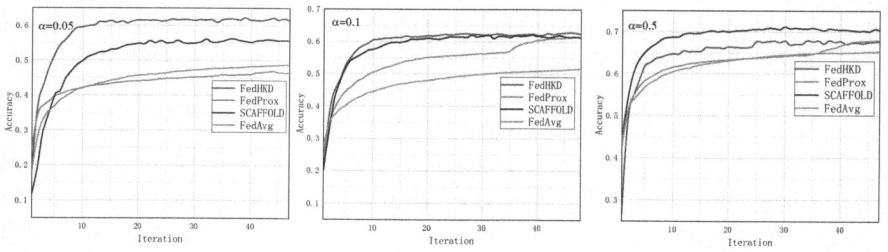

Fig. 2. Test Accuracy on the CIFAR-10 Dataset under Varying Non-IID Levels

under varying degrees of Non-IID conditions. A comprehensive analysis reveals that the FedHKD algorithm exhibits significant advantages when processing data with different levels of Non-IID data (α=0.1, 0.5, 0.05). First, when α=0.1 and 0.5, although the loss value of FedHKD is slightly higher, its fluctuations are minimal, demonstrating good stability. This indicates that FedHKD can maintain consistent performance even under higher Non-IID conditions. Secondly, when α=0.05, the loss value of FedHKD is markedly lower than that of the other algorithms, showcasing its exceptional performance in a highly Non-IID environment. Overall, the FedHKD algorithm performs well across various degrees of Non-IID data, particularly in highly Non-IID scenarios, where its lower loss value highlights its excellent adaptability and stability.

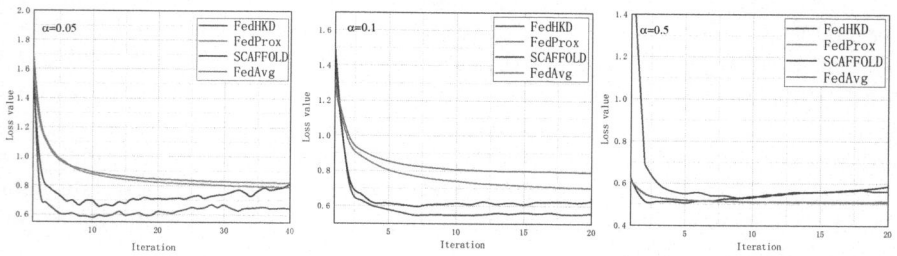

Fig. 3. Loss value of Fashion-MNIST Dataset under Varying Non-IID Levels

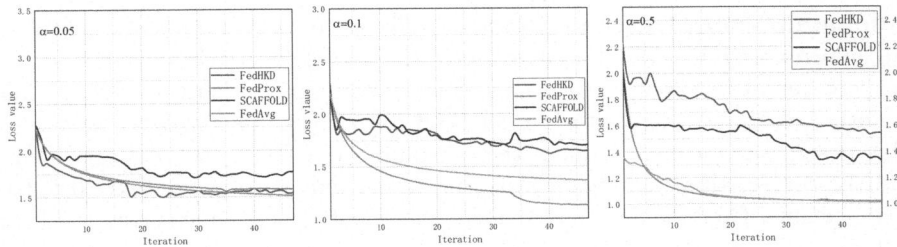

Fig. 4. Loss value of CIFAR-10 Dataset under Varying Non-IID Levels

Knowledge transfer between clusters: As shown in Table 1, knowledge distillation enhances the performance of the global model on the Fashion-MNIST and CIFAR-10 datasets, particularly under highly Non-IID conditions. By implementing knowledge distillation, the accuracy, F1 score and recall have all seen significant improvements. For instance, on the Fashion-MNIST dataset, when the degree of Non-IID is high (α=0.05), the accuracy increases from 70.62% to 79.12%, and the CIFAR-10 dataset, accuracy rises from 46.46% to 58.65% in CIFAR-10. At the same time, there are notable improvements in both the F1 score and recall rate. This indicates that knowledge distillation effectively enhances the generalization ability of the global model by facilitating knowledge sharing between clusters, particularly in highly Non-IID scenarios. As a result, the global model can classify more accurately and significantly reduce misclassifications.

Table 1. Comparison of FedHKD(NKD) and FedHKD on Fashion-MNIST and CIFAR10 datasets. FedHKD (NKD) is the performance without knowledge distillation

Metric	Non-IID Degree	Fashion-MNIST		CIFAR-10	
		FedHKD(NKD)	FedHKD	FedHKD(NKD)	FedHKD
Accuracy	α=0.05	70.62%	**79.12%**	46.46%	**58.65%**
	α=0.1	70.49%	**79.04%**	55.03%	**59.37%**
	α=0.5	77.96%	**81.05%**	62.41%	**64.53%**
F1 Score	α=0.05	68.86%	**78.16%**	41.05%	**58.37%**
	α=0.1	69.91%	**78.29%**	53.73%	**59.08%**
	α=0.5	77.85%	**80.63%**	61.47%	**64.34%**
Recall	α=0.05	70.62%	**78.76%**	46.46%	**58.65%**
	α=0.1	70.49%	**79.04%**	55.05%	**59.37%**
	α=0.5	77.92%	**80.94%**	62.41%	**64.76%**

Communication Overhead Between Client and Server: Compared to the FedProx and SCAFFOLD algorithms, the FedHKD algorithm requires only the representative client in each cluster to communicate with the server, while the other clients in the cluster do not need to interact directly with the server. This design significantly reduces the communication overhead between clients and the server, allowing devices with limited resources to participate in federated learning. Consequently, it enhances the scalability and inclusiveness of the system.

6 Conclusions

This paper presents FedHKD, a hierarchical federated learning algorithm integrating clustering and knowledge distillation to tackle Non-IID data challenges. By grouping clients with similar data distributions and using knowledge transfer, FedHKD improves model performance while reducing communication costs. Experiments demonstrate significant accuracy gains, up to 14.24%, with reduced communication overhead. Future work will explore optimizing cluster leader selection and extending the algorithm to real-world applications.

Acknowledgements. This work is partially supported by the Humanities and Social Sciences Fund of the Ministry of Education (No. 24YJAZH150 and No. 22YJCZH106), the Guangdong Provincial Regional Joint Fund (No. 2022A1515110980), and the Key Project of Natural Science Research of Guangdong Provincial Department of Education (No. 2020ZDZX3060).

Appendix

FedHKD Convergence Analysis

Assume that each cluster has an optimal parameter set θ^*, which minimizes the loss function within that cluster. After clustering, the mean (μ), standard deviation (σ), and skewness (γ) of the data distributions among the clients within the same cluster become more similar. This similarity results in the gradient updates from these clients having more consistent directions and magnitudes, effectively reducing the variance in parameter updates within the cluster. Consequently, the local updates performed by each client within the cluster are more likely to align their model parameters θ_i^t with the cluster's optimal parameters θ^*, rather than drifting toward the average parameters $\bar{\theta}_t$ across clusters.

For each client i parameter θ_i^t, it is defined to be within the δ-neighborhood of the optimal parameter θ^*, that is:

$$\left\| \theta_i^0 - \theta^* \right\| \leq \delta \tag{5}$$

where, δ is a small constant, indicating that the difference between the client model parameters can be very close to the optimal value

In the same cluster, the gradients $\nabla \ell(\theta)$ of the clients are very similar, indicating that their loss functions $\ell(\theta)$ are converging along similar trajectories. This leads to more coordinated local updates and facilitates the convergence of the overall model parameters toward the cluster's optimal parameters θ^*. We can measure the convergence rate by

$$v_t = \min_{c \in [C]} \left\{ \eta_t \lambda_{\min} \left(\nabla^2 \ell(\theta_t) \right) \right\} \tag{6}$$

$\lambda_{\min}(\cdot)$ is the minimum eigenvalue of the Hessian matrix $\nabla^2 \ell(\theta_t)$ and reflects the curvature of the loss function. The proper step size toward the minimum can be measured by the convergence factor v_t. The similarity of models within a cluster means that the curvature of the loss function within the cluster (reflected by the minimum eigenvalue $\lambda_{\min}$ of the Hessian matrix) is more stable.

To derive the expected difference between the current model parameters and the optimal parameters, we can start with the basic model update formula:

$$\theta_{t+1} - \theta^* = \theta_t - \theta^* - \eta_t \nabla F(\theta_t) \tag{7}$$

Square both sides:

$$\|\theta_{t+1} - \theta^*\|^2 = \|\theta_t - \theta^*\|^2 - 2\eta_t(\theta_t - \theta^*)^T \nabla F(\theta_t) + \eta_t^2 \|\nabla F(\theta_t)\|^2 \tag{8}$$

After taking the expectation, using the convexity assumption and the boundedness of the gradient, we get:

$$E[\|\theta_{t+1} - \theta^*\|^2] \leq (1 - v_t)E[\|\theta_t - \theta^*\|^2] + \eta_t^2 \sigma^2 \tag{9}$$

Introducing the cumulative term, we finally get:

$$E[\|\theta_t - \theta^*\|^2] \leq (1 - v_t)^t \|\theta_0 - \theta^*\|^2 + \frac{\sigma^2}{2\mu} \sum_{\tau=1}^{t} (1 - v_\tau)^{t-\tau} \eta_\tau^2 \tag{10}$$

where σ^2 is the variance of the stochastic gradient and μ is a strong convexity constant. The term $\frac{\sigma^2}{2\mu} \sum_{\tau=1}^{t} (1 - v_\tau)^{t-\tau} \eta_\tau^2$ quantifies the accumulation of noise in this gradient change over multiple training rounds in the federated learning process.

Homogeneity in the cluster can effectively reduce the fluctuation of gradient variance σ^2. By reducing gradient variance σ^2, the negative impact of gradient fluctuations on model training can be alleviated, ultimately leading to faster and more stable model convergence faster and more stable.

References

1. McMahan, B., Moore, E., Ramage, D., Hampson, S. and y Arcas, B.A.: Communication-efficient learning of deep networks from decentralized data. In Artificial intelligence and statistics, pp. 1273–1282 (2017)

2. Ma, X., Zhu, J., Lin, Z., Chen, S., Qin, Y.: A state-of-the-art survey on solving non-iid data in federated learning. Futur. Gener. Comput. Syst. **135**, 244–258 (2022)
3. Li, T., Sahu, A.K., Zaheer, M., Sanjabi, M., Talwalkar, A., Smith, V.: Federated optimization in heterogeneous networks. In: Proceedings of Machine Learning and Systems **2**, 429–450 (2020)
4. Karimireddy, S.P., Kale, S., Mohri, M., Reddi, S., Stich, S., Suresh, A.T.: Scaffold: Stochastic controlled averaging for federated learning. In: International Conference on Machine Learning, pp. 5132–5143. PMLR (2020)
5. Wang, J., Liu, Q., Liang, H., Joshi, G. and Poor, H.V. Tackling the objective inconsistency problem in heterogeneous federated optimization. Adv. Neural Inform. Proc. Syst. **33**, 7611–7623 (2020)
6. Zhu, H., Jinjin, X., Liu, S., Jin, Y.: Federated learning on non-iid data: A survey. Neurocomputing **465**, 371–390 (2021)
7. Yang, L., Huang, J., Lin, W., Cao, J.: Personalized federated learning on non-iid data via group-based meta-learning. ACM Trans. Knowl. Discov. Data **17**(4), 1–20 (2023)
8. Ma, X., Zhang, J., Guo, S., Xu, W.: Layer-wised model aggregation for personalized federated learning. In Proceedings of the IEEE/CVF Conference on Computer Vision And Pattern Recognition, pp. 10092–10101 (2022)
9. Gou, J., Yu, B., Maybank, S.J., Tao, D.: Knowledge distillation: A survey. Int. J. Comput. Vision **129**(6), 1789–1819 (2021)
10. Brophy, E., Wang, Z., She, Q., Ward, T.: Generative adversarial networks in time series: A systematic literature review. ACM Comput. Surv. **55**(10), 1–31 (2023)
11. Rousseeuw, P.R.: Silhouettes: a graphical aid to the interpretation and validation of cluster analysis. J. Comput. Appl. Math. **20**, 53–65 (1987)
12. Caliński, T., Harabasz, J.: A dendrite method for cluster analysis. Commun. Stat.-Theory Methods **3**(1), 1–27 (1974)
13. Davies, D., Bouldin, D.: A cluster separation measure: IEEE transactions on pattern analysis and machine intelligence. itpidj 0162-8828, pami-1, 2 224–227. Crossref Web of Science (1979)

Application of Ensemble Learning Based on High-Dimensional Features in Financial Big Data

Yexin Zhang[1] , Yunhao Li[1] , Gaoming Zhang[1], Ziyu Ding[1] , Yaqi Wu[1] ,
and Yun Peng[1,2]( )

[1] Institute of Artificial Intelligence, Guangzhou University, 510006 Guangzhou, China
yexinzhang@e.gzhu.edu.cn, {yunpeng,yunpeng}@gzhu.edu.cn
[2] Guangdong Provincial Key Laboratory of Blockchain Security, 510006 Guangzhou,
China

Abstract. Currently, most of the research on Internet financial models is based on traditional machine learning models, which often fail to adequately capture the complex features and potential non-linear interactions within financial data. In addition, the training process of the models is usually complex and time-consuming, making it difficult to meet the demand for rapid processing and analysis of financial data. To address these issues, this paper proposes a new approach that combines high-dimensional feature engineering with ensemble learning. Unlike most existing feature engineering methods, this paper specifically focuses on the high-dimensional features of financial data, and thoroughly investigates how to efficiently construct and select more useful features to fully explore the potential information of the data. We combine tree-based ensemble methods (e.g., LightGBM and XGBoost) with neural network models to construct a multi-level ensemble learning framework for complex financial data features. Experimental results show that the method not only generates high-quality and informative features, but also significantly enhances the robustness and prediction accuracy of the model, effectively solving the challenges for modeling and training complex financial data, especially when facing large-scale and multi-dimensional datasets.

Keywords: Financial Big Data · High-Dimensional Feature
Engineering · Ensemble Learning

This work was supported in part by the Guangzhou Basic and Applied Basic Research Project SL2024A03J00397, the National Natural Science Foundation of China (NSFC) under Grant 62472116, and the Natural Science Foundation of Guangdong Province under Grant 2023A1515030273.

F. Zhang et al. (Eds.): AIS&P 2024, LNCS 15399, pp. 117–130, 2025.
https://doi.org/10.1007/978-981-96-1148-5_10

1   Introduction

Datasets with multiple features are known as high-dimensional data [1], and high-dimensional feature engineering, which involves the creation and selection of relevant features from this data, plays a crucial role in financial data analytics, especially given the exponential growth of domain data. High-dimensional data is often complex and sparse, and the direct application of traditional machine learning algorithms may lead to the curse of dimensionality. Therefore, developing effective feature extraction methods becomes crucial. Due to the large search space associated with high-dimensional data, extracting relevant and valuable knowledge solely from feature extraction is challenging, making feature selection indispensable [2]. Feature selection is the process of removing features with less informative content and selecting those with greater discriminative power to better represent the data [3]. Higher-dimensional data is more challenging for feature selection than lower-dimensional data because, due to the curse of dimensionality, the decision regarding which features to select is more complex. Furthermore, the larger number of features necessitates consideration of computational efficiency during selection [4].

A single predictive model may struggle to adapt to changes, leading to mismatches that result in inaccurate and unstable predictions [5]. Ensemble learning produces more accurate predictions by combining the outputs of multiple machine learning models. Its key advantage lies in enhancing the performance of individual algorithms by creating a more robust solution through model combination [6]. However, traditional machine learning models often fail to capture the complex characteristics of financial data, such as non-smoothness, non-linearity, serial correlation, and non-linear interactions. Current models like Support Vector Machines (SVM), Random Forests (RF), and Artificial Neural Networks (ANN) are designed to model complex, primarily non-linear relationships in financial data. While these models address non-linearities, they often overlook dependencies between successive data points [7]. Although most supervised learning processes in ensemble frameworks are automated, few can robustly acquire raw data and generate high-quality predictions without user intervention or software errors [8]. Ensemble learning currently focuses on algorithm selection and hyperparameter optimization, a process that requires significant computational resources and time to explore a vast search space.

Based on the above discussion and problem analysis, this paper proposes a large-scale ensemble learning framework for financial big data platforms, combining high-dimensional feature engineering and ensemble learning techniques. During feature extraction, we first integrate multi-dimensional data to construct a rich feature set, enhancing the effectiveness of the information. Ensemble learning is a model that deploys multiple learning schemes instead of relying on a single algorithm to solve the same problem. This multi-level prediction scheme improves performance by training multiple models and combining their predictions [9]. Consequently, model training employs an ensemble learning approach to enhance accuracy. Diversity among base learners is an essential factor when implementing ensemble learning classifiers [10]. This paper details the algorithms employed and the specific methods of feature engineering. Finally, the paper presents the results of model experiments to validate the proposed method's effectiveness and compares it with existing techniques. These results not only confirm the superiority of our method but also provide valuable references for future research.

The main contributions of this paper are as follows:

1. We propose a customised high-dimensional feature processing strategy that integrates multi-dimensional data and constructs expressive feature sets to address the complexity, sparsity, and volume of financial data. This strategy improves the model's generalisation ability through efficient feature selection, mitigating the curse of dimensionality and reducing redundancy, marking a significant advancement in financial big data processing.
2. We develop a large-scale ensemble learning training framework specifically for financial big data platforms, integrating high-dimensional feature engineering with ensemble learning for the first time. This framework employs a multi-layer stack integration strategy for model training, enhancing accuracy while minimising user expertise requirements and human intervention via automated data preprocessing, model selection, and hyperparameter tuning, an area not thoroughly explored in existing research.
3. Existing studies often focus on model selection and hyperparameter optimisation, neglecting feature engineering's role in improving performance. We enhance predictive performance and stability by combining accurate feature selection, extraction, and construction with an ensemble learning framework. Experimental results show that our framework achieves high prediction accuracy on raw financial datasets, significantly reducing development time and costs, representing a notable innovation compared to existing techniques.

2   Related Work

2.1   Feature Engineering

Feature engineering is a key step in the field of data science and machine learning, aiming to transform raw data into more representative features to improve model performance and predictive accuracy [11]. Feature engineering for financial big data, which involves extracting, processing, and optimizing features from large amounts of raw financial data, has become crucial in financial big data analytics and plays a vital role in risk management, customer segmentation, and fraud detection [12,13]. The purpose of feature engineering is to transform raw data into a form that can be understood and utilized by models. This process typically involves feature selection and feature extraction, both of which are data preprocessing techniques used to enhance the quality of the feature space [14].

Feature extraction involves applying linear or non-linear transformations to the original data to reduce its dimensionality [15]. This process often incorporates feature construction, which creates new, higher-level features from the original ones to reduce dimensionality and improve classification performance [16].

Feature extraction has been widely used in the financial domain as a fundamental step to improve model performance. For example, in credit scoring models, constructing features such as income-to-debt ratios and credit card usage can significantly enhance predictive power [17,18].

In high-dimensional spaces, as the number of dimensions increases, not only does the complexity of the dataset grow, but the number of uninformative features associated with class concepts also rises due to irrelevance and redundancy [19]. Feature selection involves identifying the smallest subset of features from the existing feature set based on specific criteria [20]. Feature selection methods can be categorized into filter, wrapper, and embedded methods, each aiming to assess and select the most representative features by evaluating their importance [21,22]. While wrapper methods select a subset of features based on classifier feedback, they tend to be biased towards the specific classifier used, leading to high computational complexity and time consumption [23]. Embedded methods, although computationally efficient, produce results that depend on the specific learning algorithm, limiting their interpretability. On the other hand, filter methods, which select features as a preprocessing step by evaluating each subset based on the intrinsic properties of the data, are computationally fast, scalable to high-dimensional datasets, and independent of the learning algorithm, meaning that feature selection is performed only once for a given training dataset [24,25].

Conventional methods often rely on expert experience to manually generate new features using mathematical and logical operations, which can overlook potentially important patterns in the data. This approach limits diversity and innovation in feature construction. Furthermore, when dealing with high-dimensional feature selection or complex data involving non-linear relationships, manual extraction may result in information loss and degrade model performance. To address these limitations, this paper employs automated feature engineering tools and deep learning techniques to mine potentially significant patterns in the data through an automated feature generation process.

2.2   Ensemble Learning

Ensemble learning [26,27] is a machine learning technique that enhances overall model performance by combining the prediction results of multiple base learners. In recent years, with the breakthroughs of deep learning in various downstream tasks, deep learning-based ensemble learning models have gained significant attention [28,29].

In the financial sector, ensemble learning has demonstrated a wide range of applications. Wang et al. [30] proposed a method that combines deep neural networks with traditional ensemble learning, showing that deep ensemble models perform effectively in handling complex data, especially in fields such as image recognition and natural language processing. Li et al. [31] explored the application of ensemble learning in financial risk prediction and asset management, with results indicating that ensemble learning models outperform single models in terms of both prediction accuracy and stability, which is critical for high-risk management scenarios. Wang et al. [32] introduced a stacking-based financial risk approval model that leverages deep learning techniques to handle imbalanced data, improving the joint loan approval rate by up to 6% through optimised feature extraction and counterfactual enhancement methods. Dasari et al. [33] developed an ensemble learning model combining bagging and voting classifier techniques to enhance loan eligibility prediction accuracy, achieving an improvement in model performance from 80% to 94%. Nguyen et al. [34] significantly improved AUC and F1 scores for churn and purchasing behaviour prediction by integrating Random Forest, CNN, and Boosting algorithms, and further optimising classifier weights using evolutionary algorithms.

However, these existing methods are complex in design and often lack sufficient attention to model interpretability, particularly in fields such as finance. In traditional ensemble learning frameworks, several tasks must be manually configured by the user, including data processing, feature selection, model discovery, model training, hyperparameter optimisation, and model deployment [35]. The concept of automated prediction has long attracted interest from statistics and machine learning researchers [36]. Additionally, as these models are trained and evaluated using historical data, data distribution tends to be highly imbalanced, and the performance metrics have yet to reach an ideal state. Consequently, the models may struggle to adapt to newly emerged data in practical applications, leading to inaccurate prediction results. Considering the complexity, this paper leverages ensemble learning algorithms to automate time-consuming manual steps, allowing users to directly use raw data for model training while modularising the entire framework to enhance scalability and interpretability.

3 Methodology

This paper proposes a large-scale ensemble learning training framework for financial big data platforms, which tightly integrates high-dimensional feature engineering with ensemble learning techniques. First, leveraging high-dimensional feature engineering, the original features are extended and refined through feature importance screening, providing a solid foundation for subsequent model training. For model training, a multi-layer stacking integration strategy is employed, combining the advantages of automation technology and machine learning to effectively enhance the model's prediction accuracy. Additionally, the framework automates model selection and hyperparameter optimisation, significantly reducing the complexity and workload associated with model construction. Finally, this paper elaborates on two key modules of the framework.

3.1 High-Dimensional Feature Engineering

Financial data is predominantly in the form of time series, and there are numerous data sources. A key focus of this section is how to extract more effective features from these basic features to improve model training. The following will explain in detail how this paper extracts, selects, and generates new features based on existing ones.

Feature Extraction. The feature extraction methods used in this section fall into two main categories: the first involves constructing new features based on existing ones, and the second involves combining existing features to create combined features. Newly constructed features are generated through classification, statistical, and behavioural analysis, while combined features further enrich the model's input via feature derivation. Both approaches aim to enhance the model's ability to understand and predict customer behaviours and needs. We employed One-Hot encoding, target encoding, and frequency encoding for categorical feature construction, generating new feature combinations by processing categorical variables. Behavioural analysis features were also extracted to uncover potential behavioural patterns through user behaviour path analysis, frequency and temporal features, and behavioural trend features. Additionally, discretisation transforms continuous variables into categorical features, which improves model interpretability.

Feature Selection. In this paper, we use correlation coefficients to analyze the linear relationship between individual features and the target variable, screening out features that are highly correlated with the target variable by measuring the degree of linear correlation between the features and the target variable. Subsequently, we adopt a feature screening method based on feature importance using an ensemble learning model (e.g., Random Forest) to further assess the importance of features. The Random Forest model calculates the importance score of a feature by evaluating the contribution of each feature to the target variable at the time of tree node splitting during the construction of multiple decision trees. The feature importance score reflects the impact of each feature on model performance during the prediction process; thus, features with higher scores are retained for subsequent model training, while those with lower scores are eliminated. This approach not only captures the non-linear relationships between features and target variables but also effectively addresses the problem of multicollinearity among features, thereby improving the generalization ability of the model.

Feature engineering in financial big data is crucial for data analysis. After conducting high-dimensional feature engineering, this paper not only enriches the feature representation of the dataset but also provides more representative features for the further model training within the framework, which enhances the performance and prediction accuracy of the model. Furthermore, by combining correlation coefficient analysis with the feature importance-based filtering method, we achieve a balance of both efficiency and effectiveness in the feature selection process, providing a more robust input feature set for the prediction model of financial big data platforms.

3.2 Ensemble Learning for Financial Data

In this paper, we construct an automated learning framework based on the idea of ensemble learning, which primarily includes data preprocessing and dataset screening, model training, and evaluation. Preprocessing is performed on financial data to identify the type of problem to be predicted (binary classification, multi-class classification, or regression). Dataset filtering is employed to divide the data into separate parts for model training and validation, fitting various models, and finally creating an optimized model ensemble that outperforms any individual trained model. The model ensemble is fitted and evaluated using cross-validation, after which the selected models undergo parameter tuning through techniques such as grid search and Bayesian optimization. The entire learning framework is described in detail in this section.

Data Preprocessing and Dataset Selection. Since financial data often contains issues such as missing values and outliers, we clean the data through missing value imputation, outlier detection, and data normalization to ensure the quality of the underlying data for feature construction. Additionally, since multiple models are used for ensemble learning, it is essential to ensure the consistency of model inputs; thus, the input data for the models need to be processed before model training. The features obtained after feature engineering in the previous section will be classified into numerical, categorical, textual, or date/time features, and then processed based on their type and the specific model. The processing methods include data normalization and standardization, textual feature encoding, temporal data type conversion, filling in missing data, removing duplicates, and outlier detection.

When training a model, an unbalanced dataset, characterized by an unequal number of positive and negative samples, can hinder the model's ability to comprehensively learn the features of the minority class during training. This may lead to a tendency to predict the majority class while neglecting the features and importance of the minority class, resulting in inaccurate predictions and poor model performance. To address the problem of dataset imbalance, we adopt an approach that screens positive and negative samples in the training set and trains the model according to a specific ratio during the training phase, while no screening of positive and negative samples is performed on the test set in the testing phase. Specifically, in the training phase, we use undersampling to dynamically adjust the sample proportion, reducing the number of samples in the majority class by randomly deleting some samples, thus making it closer to the number of samples in the minority class. Undersampling effectively reduces the proportion of samples in the majority class, allowing the model to better focus on the minority class samples. In the testing phase, we maintain the initially set proportion to preserve the natural distribution of the dataset.

There are two benefits to adopting this method: first, by screening positive and negative samples in the training set during the training phase, we can achieve a more balanced proportion of samples, avoiding the model being affected by data skewness. This effectively reduces the risk of overfitting due to unbalanced samples, enhances the model's stability, and enables better learning of the features of both positive and negative samples, thus improving the accuracy and robustness of predictions. Second, by not screening the test set for positive and negative samples during the testing phase and maintaining its natural distribution, we can more realistically reflect the model's performance in actual application scenarios. This ensures that the model will not be biased due to proportion adjustments made in the training phase, effectively assessing its performance in handling unseen data, and ensuring that the test results are objective and authentic.

Model Training and Evaluation. First, the model training process begins with data preprocessing, during which the framework automatically handles missing values in the dataset through methods such as imputation, interpolation, or other strategies to ensure data integrity. Following the feature engineering process in the previous section, which encodes categorical variables and standardizes and normalizes numerical features, the model is better equipped to understand the data. Additionally, dataset filtering enhances the model's ability to generalize across different datasets. During training, the framework automatically performs model selection and employs automated hyperparameter optimization techniques, such as Bayesian optimization, to tune the model's hyperparameters. These optimization techniques aim to identify the parameter settings that yield the best model performance by exploring various combinations of hyperparameters. Bayesian optimization enhances optimization efficiency by constructing a probabilistic model of the relationship between hyperparameters and the model's performance, intelligently selecting the next parameter combination.

Unlike traditional machine learning approaches, where only a single model is selected for training, this paper advocates for the combination of outputs from multiple models. This approach is justified by the observation that different models may exhibit unique strengths on varying data types or prediction tasks. By combining these models, the risk of overfitting can be mitigated, and the model's generalization ability can be improved. Moreover, since model prediction does not require a secondary response, multiple models can be trained for multi-model fusion. We selected a diverse array of base learners, including tree-based models (e.g., LightGBM and XGBoost) [37,38] and neural network models [39], both of which are effective in managing nonlinear relationships and are particularly suitable for sparse, high-dimensional data. To enhance the models' generalization ability, we adopt a multi-layer stacked integration strategy. In this framework, the first layer comprises multiple base learners (e.g., LightGBM, XGBoost, and multilayer perceptron), which are trained independently on the same dataset to generate preliminary prediction results. These results are then fed into a meta-learner (e.g., a linear regression model), which further fuses these predictions to produce the final output. To optimize model performance, we utilize a combination of grid search and stochastic search for hyperparameter tuning and introduce Bayesian optimization techniques to automatically identify the optimal configuration, thereby enhancing model results further.

After model training is completed, a comprehensive evaluation of the model is conducted using performance metrics such as accuracy, F1 score, and AUC (area under the curve) to assess its effectiveness. Specifically, evaluation is performed using cross-validation techniques, wherein the dataset is divided into multiple subsets, rotating through these different subsets for training and validation to obtain a robust estimate of model performance. Cross-validation helps identify variations in model performance across different data subsets, ensuring model stability and reliability. During the evaluation phase, the model's hyperparameters are readjusted to further optimize performance. By testing under various hyperparameter settings, the framework can identify the best combination of parameters to improve the final model.

4   Experiments

Based on the framework introduced in the previous methodology section, this paper validates the effectiveness of the two modules within the framework. First, we assess the multilayer stacking strategy of the integrated learning framework by comparing the prediction results and model selection of multiple models. The experimental results demonstrate that our framework accurately selects the best model and that the chosen ensemble learning framework is both effective and more efficient in terms of time. Additionally, we conduct ablation experiments on the designed modules, testing each module individually as well as their combined effect. The results indicate that both modules are effective on their own, and the prediction accuracy of the framework improves further when the two are combined.

4.1   Experimental Setup

A variety of features generated through feature engineering were utilized to construct the experimental dataset. The primary objective of the experiment was to evaluate the prediction accuracy of the integrated model. The experiment was conducted on a high-performance computing platform to ensure efficient processing and analysis of large-scale data.

- **Dataset.** The dataset used for the experiment is the 'Bank Term Deposit Subscription Dataset' [40] from the Kaggle platform. The objective of the experiment is to build machine learning models that classify whether a customer subscribes to a term deposit after receiving a call from a bank representative. This dataset, obtained from the UCI Machine Learning Repository, consists of 11 categorical features and 10 numerical features.
- **Configuration.** Processor (CPU): The experimental platform is equipped with 16 CPU cores, which support multi-threaded parallel computing and can accelerate the data processing and model training processes. Memory (RAM): The server is equipped with 32GB of RAM, providing sufficient memory space for large-scale data loading and processing, which helps improve computational efficiency and stability. Storage: High-performance storage devices are used to store datasets and experimental results, ensuring fast read and write speeds, thereby further enhancing overall computational performance.

Table 1. The results of the performance comparison of the models trained on the Financial Big Data platform are shown. The table shows the comparison of the accuracy (accuracy) of each model on the test and validation sets, as well as their prediction time and training time. These models include classical tree models (e.g. LightGBM, XGBoost and RandomForest) as well as neural network models (e.g. NeuralNetFastAI, NeuralNetTorch)

model	score_test	score_val	pred_time_test	pred_time_val	fit_time
LightGBM	0.9217	0.9178	0.0363	0.0360	1.3401
XGBoost	0.9210	0.9175	0.1320	0.1045	2.6482
LightGBMXT	0.9192	0.9173	0.0678	0.0378	1.2993
CatBoost	0.9184	0.9175	0.2906	0.0549	6.4781
LightGBMLarge	0.9178	0.9165	0.0611	0.0371	3.0588
NeuralNetFastAI	0.9178	0.9146	2.2119	0.5042	49.9605
RandomForestGini	0.9161	0.9143	0.2081	1.0177	0.8442
NeuralNetTorch	0.9158	0.9134	0.2486	0.3725	40.3666
RandomForestEntr	0.9140	0.9149	0.1695	0.9914	0.6958
ExtraTreesEntr	0.9117	0.9111	0.2404	1.1285	0.7211
ExtraTreesGini	0.9099	0.9104	0.2379	1.1424	0.7048
KNeighborsDist	0.9008	0.9010	0.0775	0.1461	0.0108
KNeighborsUnif	0.9004	0.9028	0.2161	0.1889	0.0113

4.2   Experimental Results

Unlike traditional machine learning, where only a single model is selected for training, this paper combines the outputs of multiple models. This approach is beneficial because different models may exhibit distinct advantages for various types of data or prediction tasks. By combining these models, the risk of overfitting can be reduced, and the generalization ability of the models can be improved. To verify the effectiveness of the modeling strategy proposed in this paper, contemporaneous testing is employed to compare the predictive performance of each model.

Contemporaneous testing is a method used to compare the effects of different strategies within the same time period. It offers several advantages: by conducting comparisons in the same time frame, it eliminates the interference of external environmental factors that may arise due to time changes, making the test results more reliable. Additionally, contemporaneous testing ensures that each strategy is evaluated in the same market environment and customer base, which enhances the fairness of the comparison. As shown in Table 1, the models yield similar accuracies, and the model with the best predictive performance is automatically selected during the training process. The comparison indicates that the tree-based model outperforms others in terms of accuracy and prediction time, while the neural network model may require more time for both prediction and training in certain scenarios.

According to the results in Table 2, the application of high-dimensional feature engineering increases the model's accuracy by 0.21 relative to the initial benchmark. The multi-layer stacked integration strategy adopted in this paper first selects the optimal model at each layer after conducting multiple model comparisons based on the model selection strategy introduced earlier, followed by an automated optimization strategy. This hierarchical integration of models effectively enhances the prediction accuracy. Table 2 evaluates the accuracy, ROC AUC, precision, F1 score, and recall of each method. The experimental results demonstrate that the method proposed in this paper shows significant improvements across all performance metrics compared to the baseline model, particularly in terms of accuracy and ROC AUC, thereby validating the effectiveness of the proposed approach.

Table 2. The results of the ablation experiments demonstrate the impact of the two types of improvement engineering (integration approach and feature engineering) used in this paper on the model prediction results. baseline model serves as a baseline for comparing the integration approach (module1), the feature engineering (module2) and its combined model (module1+module2), as well as the final improved model (ours)

method	accuracy	roc_auc	precision	f1	recall
baseline	0.7037	0.7145	0.2360	0.6027	0.7284
+integrated approach(module1)	0.8902	0.8941	0.6323	0.4424	0.3402
+feature engineering(module2)	0.9163	0.9485	0.6609	0.5944	0.5401
+module1+module2	0.9192	0.9496	0.6629	0.5866	0.5261
ours	**0.9217**	**0.9503**	**0.6844**	**0.6252**	**0.5754**

In the experiments, to further enhance model performance, the bagging integration method is introduced, which divides the dataset into multiple disjoint subsets using k-fold cross-validation (k = 5 in this paper). Each subset serves as "Out-Of-Fold" (OOF) data during model training, meaning that each model uses all data except for the current subset during training. This approach allows each model to learn from different perspectives of the data, reducing its dependence on any specific dataset and effectively mitigating the risk of overfitting. However, even k-fold cross-validation may be susceptible to overfitting, especially when the model training process overly relies on OOF data. To address this issue, this paper employs repeated bagging, which involves performing k-fold cross-validation multiple times during the training process and averaging the obtained OOF predictions to reduce the variance of the prediction results. The results presented in Table 2 demonstrate that by adopting this method, the accuracy of the model has increased by 0.25.

Overall, the results of the inter-period tests indicate that the model performs more stably across different time periods, exhibiting better generalization ability and robustness. Through further analysis and optimization, the performance of the model in each time period can be improved to ensure its reliability in practical applications.

5   Summary

In this paper, by combining high-dimensional feature engineering and ensemble learning techniques, we explore how to extract useful information from massive and complex financial data and construct efficient prediction models. The method of combining feature engineering and ensemble learning adopted in this paper has multiple advantages. By deeply mining and screening the complex features of financial data through basic feature construction, derived feature generation, and feature selection techniques, we significantly reduce redundant features, improve training efficiency, and mitigate the impact of the curse of dimensionality. Meanwhile, through the multi-model fusion ensemble learning framework, we leverage the advantages of different models to significantly improve prediction accuracy and model robustness, especially when dealing with complex, nonlinear, and sparse financial big data. Experimental results show that this integrated framework not only generates high-quality features but also significantly enhances predictive performance and stability while drastically reducing model training time and achieving robustness and efficiency that cannot be matched by existing techniques.

6   Limitations

Although this paper validates the effectiveness of the method through multiple datasets, the generalization ability of the model in different financial environments still needs further exploration. Future research should extend the experimental scope to include more financial datasets with varying feature distributions and market volatility, especially data from different economic cycles and regions, to test the stability and robustness of the model. Additionally, transfer learning techniques can be introduced to improve the generalization ability of the model across multiple environments, and dynamic testing can be conducted alongside real-time data to observe the model's performance in rapidly changing market conditions. These improvements will contribute to a more comprehensive assessment of the model's reliability for real-world applications.

References

1. Jia, W., Sun, M., Lian, J., Hou, S.: Feature dimensionality reduction: a review. Complex Intell. Syst. **8**(3), 2663–2693 (2022)
2. Pes, B.: Ensemble feature selection for high-dimensional data: a stability analysis across multiple domains. Neural Comput. Appl. **32**(10), 5951–5973 (2020)
3. Solorio-Fernández, S., Carrasco-Ochoa, J.A., Martínez-Trinidad, J.F.: A new hybrid filter–wrapper feature selection method for clustering based on ranking. Neurocomputing **214**, 866–880 (2016)

4. Bommert, A., Sun, X., Bischl, B., Rahnenführer, J., Lang, M.: Benchmark for filter methods for feature selection in high-dimensional classification data. Comput. Stat. Data Anal. **143**, 106839 (2020)
5. Wang, F., Li, Y., Liao, F., Yan, H.: An ensemble learning based prediction strategy for dynamic multi-objective optimization. Appl. Soft Comput. **96**, 106592 (2020)
6. Shankar, A., et al.: An intelligent recommendation system in e-commerce using ensemble learning. Multimedia Tools Appl. **83**(16), 48521–48537 (2024)
7. Jose, D.M., Vincent, A.M., Dwarakish, G.S.: Improving multiple model ensemble predictions of daily precipitation and temperature through machine learning techniques. Sci. Rep. **12**(1), 4678 (2022)
8. Erickson, N., et al.: AutoGluon-Tabular: robust and accurate AutoML for structured data. arXiv preprint arXiv:2003.06505 (2020)
9. Kumar, M., Bajaj, K., Sharma, B., Narang, S.: A comparative performance assessment of optimized multilevel ensemble learning model with existing classifier models. Big Data **10**(5), 371–387 (2022). https://doi.org/10.1089/big.2021.0257
10. Mienye, I.D., Sun, Y.: A survey of ensemble learning: concepts, algorithms, applications, and prospects. IEEE Access **10**, 99129–99149 (2022). https://doi.org/10.1109/ACCESS.2022.3207287
11. Dong, G., Liu. H.: Feature Engineering for Machine Learning and Data Analytics. CRC Press (2018)
12. Fritsch, C., Abt, R., Renz, B.: The benefits of a feature model in banking. In: Proceedings of the 24th ACM Conference on Systems and Software Product Line, vol. A, pp. 1–11 (2020)
13. Khadivizand, S., et al.: Towards intelligent feature engineering for risk-based customer segmentation in banking. In: Proceedings of the 18th International Conference on Advances in Mobile Computing and Multimedia, pp. 74–83 (2020)
14. Tran, B., Xue, B., Zhang, M.: Genetic programming for feature construction and selection in classification on high-dimensional data. Memetic Comput. **8**, 3–15 (2016)
15. Bolón-Canedo, V., Sánchez-Marono, N., Alonso-Betanzos, A., Benítez, J.M., Herrera, F.: A review of microarray datasets and applied feature selection methods. Info. Sci. **282**, 111–135 (2014)
16. Tran, B.N.: Evolutionary computation for feature manipulation in classification on high-dimensional data, PhD thesis, Open Access Te Herenga Waka-Victoria University of Wellington (2018)
17. Baesens, B., Roesch, D., Scheule, H.: Profit-driven data mining for direct marketing: a case study. J. Risk Financ. **17**(2), 252–268 (2016)
18. Abedin, M.Z., Hajek, P., Sharif, T., Satu, M.S., Khan, M.I.: Modelling bank customer behaviour using feature engineering and classification techniques. Res. Int. Bus. Fin. **65**, 101913 (2023)
19. Vergara, J.R., Estévez, P.A.: A review of feature selection methods based on mutual information. Neural Comput. Appl. **24**, 175–186 (2014)
20. Bolón-Canedo, V., Sánchez-Maroño, N., Alonso-Betanzos, A.: Feature selection for high-dimensional data. Prog. Artif. Intell. **5**, 65–75 (2016)
21. Guyon, I., Elisseeff, A.: An introduction to variable and feature selection. J. Mach. Learn. Res. **3**, 1157–1182 (2003)
22. Ben Jabeur, S., Stef, N., Carmona, P.: Bankruptcy prediction using the XGBoost algorithm and variable importance feature engineering. Comput. Econ. **61**(2), 715–741 (2023)
23. Mohammad Ahmadi Ganjei and Reza Boostani: A hybrid feature selection scheme for high-dimensional data. Eng. Appl. Artif. Intell. **113**, 104894 (2022)

24. Saeys, Y., Inza, I., Larranaga, P.: A review of feature selection techniques in bioinformatics. Bioinformatics **23**(19), 2507–2517 (2007)
25. García-Torres, M., Gómez-Vela, F., Melián-Batista, B., Moreno-Vega, J.M.: High-dimensional feature selection via feature grouping: a variable neighborhood search approach. Info. Sci. **326**, 102–118 (2016)
26. Mohammed, A., Kora, R.: A comprehensive review on ensemble deep learning: opportunities and challenges. J. King Saud Univ. Comput. Info. Sci. **35**(2), 757–774 (2023)
27. Zhang, C., et al.: Prefer: prompt ensemble learning via feedback-reflect-refine. Proc. AAAI Con. Artif. Intell. **38**, 19525–19532 (2024)
28. Yang, Y., Lv, H., Chen, N.: A survey on ensemble learning under the era of deep learning. Artif. Intell. Rev. **56**(6), 5545–5589 (2023)
29. Ganaie, M.A., Hu, M., Malik, A.K., Tanveer, M., Suganthan, P.N.: Ensemble deep learning: a review. Eng. Appl. Artif. Intell. **115**, 105151 (2022)
30. Wang, T., Sun, J., Zhao, X.: Deep ensemble learning for complex data processing. J. Artif. Intell. Res. **18**(4), 345–362 (2023)
31. Li, X., Zhang, H., Chen, Y.: Applications of ensemble learning in financial prediction. J. Fin. Technol. **9**(1), 67–82 (2021)
32. Wang, Y., Wang, M., Pan, Y., Chen, J.: Joint loan risk prediction based on deep learning-optimized stacking model. Eng. Rep. **6**(4), e12748 (2024)
33. Dasari, Y., Rishitha, K., Gandhi, O.: Prediction of bank loan status using machine learning algorithms. Int. J. Comput. Digit. Syst. **14**(1), 1–1 (2023)
34. Nguyen, T.M., Le, T.A., Nguyen, T.H.: A flexible framework for customer behavior prediction based on ensemble learning. In: Proceedings of the 12th International Symposium on Information and Communication Technology, pp. 126–134 (2023)
35. Oliveira, S.D., Topsakal, O., Toker, O.: Benchmarking automated machine learning (AutoML) frameworks for object detection. Information **15**(1) (2024)
36. Shchur, O. et al.: AutoGluon–TimeSeries: AutoML for probabilistic time series forecasting. In: International Conference on Automated Machine Learning, pp. 9–1. PMLR (2023)
37. Ke, G., et al.: LightGBM: a highly efficient gradient boosting decision tree. Adv. Neural Info. Process. Syst. **30** (2017)
38. Chen, T., Guestrin, C.: XGBoost: a scalable tree boosting system. In: Proceedings of the 22nd ACM SIGKDD International Conference on Knowledge Discovery and Data Mining, pp. 785–794 (2016)
39. Günther, F., Fritsch, S.: Neuralnet: training of neural networks. R J. **2**(1), 30 (2010)
40. Bank term deposit predictions. https://www.kaggle.com/datasets/thedevastator/bank-term-deposit-predictions

Collaborative Framework for Dynamic Knowledge Updating and Transparent Reasoning with Large Language Models

Ziyu Ding[1] , Pei-Gen Ye[2] , Yaqi Wu[1] , and Huali Ren[3(✉)]

[1] Institute of Artificial Intelligence, Guangzhou University, Guangzhou, China
{dingziyu,winnerwu}@e.gzhu.edu.cn
[2] School of Cyberspace Science and Technology, Beijing Institute of Technology,
Beijing, China
[3] School of Cyberspace Security, Guangzhou University, Guangzhou, China
huali.ren@e.gzhu.edu.cn

Abstract. Large-scale language models (LLMs) have made remarkable achievements in natural language processing. However, when confronted with new knowledge that is absent from their training data, they often suffer from experience issues such as inaccurate reasoning, hallucinations, and insufficient transparency in their decision-making processes. To address these challenges, we propose a framework that integrates LLMs with knowledge graphs (KGs) to enable collaborative reasoning. By employing automated prompt engineering and dynamically updating the knowledge graphs, this framework enhances reasoning accuracy. Additionally, by explicitly designing retrieval reasoning paths, the transparency and explainability of the reasoning process are significantly improved. Experimental results demonstrate that this framework achieves promising performance in knowledge graph reasoning tasks, and effectively increases the reliability of the reasoning process.

Keywords: Knowledge Graph · Large Language Model · Knowledge Update · Transparent Reasoning

1 Introduction

Large language models (LLMs) [12] have advanced rapidly due to their powerful contextual text generation capabilities, significantly accelerating progress in artificial intelligence and machine learning. The LLMs such as GPT-3 [4] and BERT [6], trained on large-scale corpora using deep learning techniques, are capable of understanding and generating natural language text, and can handle complex semantic [20] and logical reasoning tasks [15,23]. Their development has greatly enhanced language understanding and generation technologies, providing robust tools for applications like automatic question-answering and information retrieval.

© The Author(s), under exclusive license to Springer Nature Singapore Pte Ltd. 2025
F. Zhang et al. (Eds.): AIS&P 2024, LNCS 15399, pp. 131–142, 2025.
https://doi.org/10.1007/978-981-96-1148-5_11

However, despite their strong performance in general language understanding tasks, LLMs still exhibit limited flexibility, especially when confronted with knowledge absent from their training data. This limitation could cause the models to produce incorrect answers or follow unreasonable reasoning paths, sometimes leading to the phenomenon known as "hallucination" [2, 20, 27]. Specifically, when LLMs encounter new fields or specialized terminology not covered during training, their inference capabilities will degrade significantly, increasing uncertainty in the output. Furthermore, most of the existing LLM functions can be regarded as black-box models, with knowledge embedded implicitly within their parameters. This results in an opaque internal decision-making process, making it difficult for users to understand the logic behind the model's generated answers. The lack of transparency not only limits the models' applicability in critical areas such as legal adjudication and medical diagnosis but also hampers efforts to further optimize the models for improved reasoning accuracy and efficiency.

To address these challenges, domain knowledge graphs (KGs) offer an effective complementary solution. KGs store information about entities (e) and their relationships (r) using structured triples (e1, r, e2), serving as a static representation of expert knowledge within a domain. In recent years, research on integrating KGs with LLMs for joint reasoning has emerged and can be broadly categorized into two kinds of methods [1, 15, 23, 28]. The first kind refers to the semantic parsing methods, such as [23, 28], which utilize LLMs to map natural language questions into logical queries executable on knowledge graphs to retrieve answers. The second kind is the retrieval enhancement methods, such as [1, 15]. These methods first retrieve the relevant triples from the knowledge graph as background knowledge and then use the LLMs to generate the final answer.

Both approaches have distinct advantages, aiming to improve the reasoning ability and performance of question-answering systems based on KGs. However, relying solely on static, pre-built KGs may overlook new information in user queries, leading to low accuracy in reasoning results. Furthermore, these methods often treat KGs primarily as factual knowledge bases, neglecting the importance of their structural information in the reasoning process. Consequently, current systems face difficulties in directly applying LLMs to KGs in a way that ensures both the correctness of the answers and transparency in the reasoning process.

In this paper, we propose a collaborative reasoning framework that integrates large language models (LLMs) with knowledge graphs (KGs), aiming to enhance the accuracy and transparency of reasoning results. Within this framework, we assume that all input information is accurate. To address new knowledge presented in the input, we design an automatic prompt engineering method that generates domain-specific prompt words, which are then input alongside a pre-trained LLM for parsing and extracting both knowledge text and question text. Subsequently, we extract entities and relations from the knowledge text, construct fact triples, and update the knowledge graph. To address the issue of opaque or difficult-to-explain answers, we propose a reasoning method based on retrieval planning. This approach leverages LLMs and KGs to generate all possible relationship paths from the question entities to potential answers. It then retrieves valid reasoning paths from the KG and performs answer reasoning through retrieval reasoning modules. The final answer is determined through a voting mechanism, and the associated reasoning paths are shared with users. Experimental results demonstrate that the proposed framework performs exceptionally well in KG reasoning tasks and significantly improves the transparency and explainability of the reasoning process.

2   Related Work

2.1   Large Language Models

Large Language Models (LLMs) [4,5,12] are AI models based on deep learning, typically comprising billions to trillions of parameters. Leveraging multi-level attention mechanisms and representation learning techniques, these models are capable of effectively capturing contextual information and semantic associations within text, enabling them to excel in various natural language processing tasks such as text generation, language understanding, machine translation, and question-answering systems.

LLM training typically involves two key stages: pre-training and fine-tuning. During the pre-training phase [25], the model learns the structure, semantics, and contextual relationships of language through self-supervised learning on large-scale text datasets. In the fine-tuning stage, the model is supervised with task-specific labeled data to adapt it to specific applications. However, due to the immense number of parameters in large models, fully retraining or fine-tuning all parameters is impractical. As a result, a range of parameter-efficient fine-tuning techniques has emerged, such as Adapter tuning [9,30], Prefix tuning [17,26], Prompt tuning [13,18], and LoRA [10,24]. These approaches fine-tune only a small subset of newly added parameters while keeping the majority of the original model parameters fixed, significantly reducing computational costs and resource consumption while maintaining performance.

Despite the success of these techniques, LLMs remain inflexible and opaque when applied to domain-specific tasks, particularly in interactive question-answering systems [20]. LLMs often struggle to handle knowledge that is not present in their training data, which can lead to logical errors or hallucinations during the reasoning process. Moreover, these systems typically provide only final answers without disclosing the underlying reasoning process, making the results opaque, unverifiable, and potentially untrustworthy to users.

2.2   KGQA

The knowledge graph G can be viewed as a set of triples $G \in E \times R \times E$, where E represents a set of entities and R represents a set of relations. This data structure is similar to a directed graph, with each node representing an entity and each edge carrying a relation label. The triple (s,r,o) represents the fact of the relationship between subject s and object o. From a logical point of view, entities correspond to constants, relations correspond to binary predicates, and the triples in the graph correspond to the atomic facts r(s,o) obtained by applying the binary predicate r to the constants s and o [3].

In research on question-answering (QA) systems based on knowledge graphs, two primary reasoning approaches are commonly adopted: embedding-based and graph neural network (GNN)-based methods. Embedding-based reasoning [7,14,21] employs vector representation learning to map entities and relations within a knowledge graph into a vector space, where scoring functions rank entities according to their relevance to the question. However, these approaches are often limited to capturing shallow features, struggling to handle higher-order connections and complex semantics. GNN-based reasoning [8,11,22] typically follows a retrieval and reasoning framework. First, a relevant subgraph is retrieved, and GNNs are used to reason over the subgraph to identify the answer entity [29]. However, these methods face challenges in comprehending natural language, often leading to semantically imprecise answers.

In recent years, many studies have sought to improve the reasoning capabilities of large language models (LLMs) in the context of knowledge graphs to answer complex questions [20]. These studies primarily adopt two methods: semantic parsing and retrieval enhancement. Semantic parsing methods use LLMs to generate query languages for questions, which are then executed on knowledge graphs to retrieve answers [16,23,28]. However, these methods typically require fine-tuning LLMs or identifying a large number of similar examples as prompts. Additionally, generated query languages often encounter execution failures, resulting in the inability to produce answers. Retrieval enhancement methods extract relevant triples from knowledge graphs to serve as supplementary knowledge, enabling LLMs to generate final answers [1,15]. Nevertheless, these methods rely on static, pre-built knowledge graphs, which may overlook newly added information from user queries, thus diminishing the accuracy of reasoning outcomes. Furthermore, they often treat knowledge graphs as mere factual repositories, neglecting the significance of structural information in the reasoning process. Consequently, enabling LLMs to directly interact with knowledge graphs to ensure answer correctness and reasoning transparency remains a significant challenge.

3   Method

3.1   Overview

Given a natural language question q and a KG $G = (e_s, r, e_o)|e_s, e_o \in V, r \in R$, where V is a set of entities (nodes) and R is a set of relation types, the task of knowledge graph question answering (KGQA) is to find a function $F(q, G)$ that can predict the answer entity $e_a \in V$ of q on G. Since real-world knowledge is continuously evolving, missing or outdated information in the KG should be updated promptly. Usually, the latest factual information is embedded in the input questions, making it essential to extract and update the KG accordingly. Additionally, to ensure the LLM-assisted answers are correct, transparent, and verifiable, for each question q, the subject entity $e_q \in V$ involved in KG and the relationship path of the reasoning process should be given.

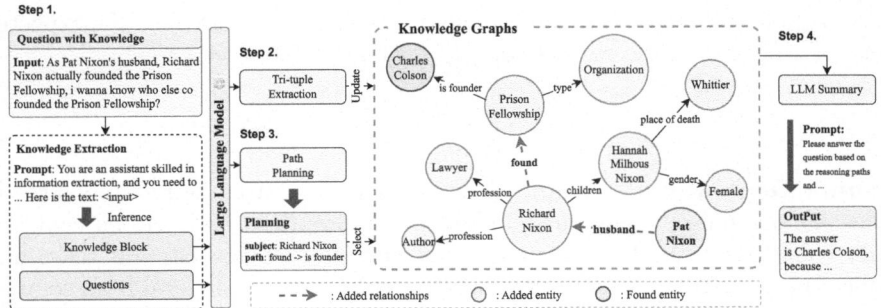

Fig. 1. The overall framework of knowledge updating and transparency reasoning

In this paper, we propose a collaborative framework for knowledge graph updating and transparent reasoning, driven by a large language model, as illustrated in Fig. 1. It consists of four main components: (1) Information extraction: extracting factual knowledge and questions from the input. (2) Knowledge update: extracting entities and relations from the factual knowledge and storing them as triples in the form <entity 1, relation, entity 2>. If these triples are missing or conflict with the KG, the KG is updated accordingly. (3) Relationship path generation: using LLM to generate possible reasoning paths for the question. (4) Retrieval reasoning: retrieving valid paths from the KG, reasoning along these paths, and outputting the answer with a transparent and verifiable reasoning process.

3.2 Information Extraction

The information extracted from complex inputs consists of two key components: factual knowledge and specific questions. Leveraging the powerful language understanding capabilities of LLMs, and to avoid the substantial computational overhead of fine-tuning large models across various domains, we adopt a more cost-efficient approach using prompt engineering. This method extracts the knowledge description text and question description text from the input X. Specifically, we focus on extracting the question text and treating the remaining input as factual knowledge.

Prompt Engineering: In this paper, to reduce the computation involved in creating and verifying effective instructions, and inspired by [31], we design an algorithm to automatically generate and select instructions using LLMs. During the automatic prompt generation process, to enhance the model's understanding, we utilize a dataset $D_{train} = (X, Y)$, which consists of K randomly selected input/output pairs from the sample set, leveraging few-shot learning techniques. The objective of the prompt is to find an instruction p such that, when the model M is prompted with the concatenation of the instruction and the input $[p; X]$, M produces the corresponding output Y. More formally, we frame this as an optimization problem, aiming to find instructions p that maximize the expected value of a per-sample score function $f(p, X, Y)$ over possible (X, Y).

$$p' = \operatorname{argmax}_p \ f(p) = \operatorname{argmax}_p \mathbb{E}_{(X,Y)}[f(p, X, Y)] \tag{1}$$

Prompt Generation: Since the search space for constructing prompts directly from the vocabulary is infinite, finding the appropriate instructions could be quite challenging. To fully leverage the knowledge of downstream tasks and reduce the computational overhead of optimizing prompt retrieval from scratch, we first enlist the help of human experts in relevant fields to establish an initial prompt word p. This initial prompt p is then input into the LLM, along with K inputs X, in a sequential manner. The BLEU score is calculated to assess the consistency between the LLM's generated output Y_{pre} and the true labeled output Y_{lab}

$$BLEU = BP * exp(\frac{1}{n}\textstyle\sum_{i=1}^{N} P_n) \tag{2}$$

BLEU is an indicator used to measure translation quality in the field of natural language processing. In the BLEU calculation, BP refers to the brevity penalty, which penalizes excessively short sentences to prevent the model from favoring shorter outputs. P_n represents the n-gram precision, and the upper limit of N is 4. The K samples are then randomly divided into N groups, and in each group, $[p, X, Y, BLEU]$ is input into the LLM to generate new prompts. Here, the BLEU score serves as a guide to direct the LLM in generating prompt words. As a result, N candidate prompts are produced.

Prompt selection: For each generated candidate prompt, a new set of K samples is randomly selected from non-overlapping datasets. The BLEU score is calculated for these subsets, and the candidate prompt with the highest BLEU score is selected as the final optimized prompt.

Prompt Selection: For each generated candidate prompt, K samples are randomly selected from non-overlapping datasets. The BLEU score is then calculated for these subsets, and the candidate prompt with the highest BLEU score is chosen as the final optimized prompt.

3.3 Knowledge Update

For the factual knowledge description text extracted from the input, we need to identify and extract the entities and relationships, when present, and construct knowledge triples in the form of <entity, relationship, entity> for the knowledge graph. To optimize the process of constructing these triples, we apply the automatic prompt engineering technique described in Sect. 3.1. The [factual knowledge, prompt2] pair is then input into the LLM in sequence, producing the extracted triples. Subsequently, SQL queries are used to check the knowledge graph. If these triples are absent from the KG or conflict with existing knowledge, the extracted triples are used to update and replace the existing knowledge in the KG.

3.4 Relationship Path Generation

To mitigate the hallucination problem in LLMs and improve the accuracy of answer generation, this paper addresses the issue by integrating an external domain knowledge graph with the LLM during pre-training in a specific domain. Additionally, to ensure that the source of the answer is credible and verifiable, the reasoning process is provided with the answer as explanation text generated by the LLM.

In the field of knowledge graphs, relation paths that capture semantic relationships between entities have been widely used across various reasoning tasks. By leveraging these relation paths, we can continuously retrieve up-to-date knowledge from the knowledge graph and support reasoning in knowledge graph question answering (KGQA) tasks. To fully harness the capabilities of language models (LLMs) with prompts, we design a simple instruction template that combines prompt words with questions to generate potential paths that can obtain the answer object via the question topic. The generated relation path takes the following form:

$$z = < PATH > r_1 < SEP > r_2 < SEP > ... < SEP > r_l < /PATH > \quad (3)$$

where <PATH>, <SEP>, </PATH> are special markers indicating the start, separator, and end of the relationship path, respectively.

3.5 Retrieval Reasoning

By the question q extracted in Sect. 3.1, the question subject e_q, the updated knowledge graph (KG) from Sect. 3.2, and the relation path z of the potential reasoning answer generated in Sect. 3.3, this section focuses on how to retrieve the reasoning path within the KG, obtain the answer, and outline the reasoning process. The answer and reasoning processes are re-input into the LLM using prompt engineering, simultaneously. To generate an answer and explanation that aligns with human reading habits. The path inferred from the question subject e_q and the relation path in the KG is described as follows:

$$W_z = w_z(e_q, e) | w_z(e_q, e) = e_q{\to}r_1{\to}e_1{\to}r_1{\to}e_2...r_l{\to}e_a, w_z(e_q, e) \in G \quad (4)$$

This paper employs a constrained breadth-first search technique to retrieve the reasoning path w_z from the knowledge graph (KG). In the experiment, all retrieved paths are utilized for reasoning. While the correct answer can be obtained through these retrieval paths, the reasoning path retrieved is not always entirely accurate, which may result in incorrect answers. However, the experiment revealed that multiple retrieval paths usually lead to correct answers in most cases. Therefore, a majority voting mechanism is applied to select the final answer.

$$a^* = \text{argmax}_{a \in A} \text{ count}(a) \tag{5}$$

where A denotes the set of all the answers output by multiple paths, and a* is the answer that appears most frequently. Finally, the answer and its corresponding reasoning path are combined with prompt4 and then input into the LLM. Finally, the LLM then generates a response, providing both the answer and an explanation, which is returned to the user for clarification.

4   Experiments

In our experiments, we aim to address the following questions: **QR1:** Can our approach effectively distinguish between questions and knowledge within complex texts? **QR2:** Compared to existing methods, can our approach demonstrate an advantage when the input includes new questions and new knowledge?

4.1   Experiment Setup

Dataset. We evaluated the performance of information extraction on the SQuAD1.1 dataset and assessed the reasoning ability of our framework on the WebQuestionsSP (WebQSP) benchmark KGQA dataset. SQuAD1.1 is a knowledge-based question-answering dataset, and we combine the questions and contexts from this dataset to serve as the evaluation data for the information extraction component. WebQSP uses Freebase, which contains approximately 88 million entities, 20,000 relations, and 1.26 billion triples, as background knowledge, and includes questions with up to 4 hops. Additionally, due to the specific nature of the research questions, we construct a supplementary dataset (QSPex) based on WebQSP, which includes 1,627 natural language descriptions of knowledge, questions, and answers, consistent with the style of WebQSP.

Baseline. We compare with the recent RoG [19] as a baseline, which is also an explainable reasoning technique that combines LLM and KG but does not consider the presence of new useful information in the input. Specifically, We compare our approach with the baseline methods across the following scenarios: using only WebQSP, using WebQSP with 300 QSPex samples, using WebQSP with 600 QSPex samples, and using WebQSP with 900 QSPex samples. The inclusion of new knowledge and questions allows us to evaluate the reasoning capability of our method when faced with inputs containing novel knowledge.

Evaluation Metrics. Following current mainstream approaches, we use Accuracy, Hits@1, F1, Precision, and Recall as evaluation metrics for the predicted answers. Additionally, we use BLEU and Hits@1 as evaluation metrics for the information extraction accuracy.

Table 1. Performance comparison for information extraction on different datasets.

Dataset	Hits@1	F1	Precision	Recall	BLEU
SQuAD	92.16	92.09	92.80	92.37	97.64
QSPex	99.74	99.74	97.61	98.33	99.94

Experiment Details. Our experiments were conducted on a server equipped with two RTX 3090-24 GB GPUs. Based on RoG [19], we used the ChatGPT-3.5-turbo model to conduct the experiments, with the temperature parameter, i.e., 0.3, to ensure output stability.

4.2 Performance Comparison

Information Extraction. Information Extraction. As observed from Table 1, our information extraction method demonstrates desirable performance on the constructed QSPex dataset and achieves high accuracy on the SQuAD1.1 dataset, proving that this method provides solid support for downstream tasks.

Table 2. Performance Comparison with Baselines after Introducing Varying Degrees of New Questions to the WebQSP Dataset.

Dataset	Methods	Hits@1	F1	Accuracy	Precision	Recall
WebQSP	RoG	85.43	49.24	80.05	46.25	80.06
	Ours	**83.64**	**46.37**	**77.56**	**43.29**	**77.56**
		(−1.79)	(−2.87)	(−2.49)	(−2.96)	(−2.50)
WebQSP+QSPex (300)	RoG	73.27	42.17	68.74	39.54	68.74
	Ours	**75.96**	**42.58**	**70.82**	**39.60**	**70.82**
		(+2.69)	(+0.41)	(+2.08)	(+0.06)	(+2.08)
WebQSP+QSPex (600)	RoG	64.89	37.38	60.96	35.03	60.96
	Ours	**70.98**	**40.55**	**66.53**	**37.60**	**66.53**
		(+6.09)	(+3.17)	(+5.57)	(+2.57)	(+5.57)
WebQSP+QSPex (900)	RoG	58.41	33.74	54.95	31.59	54.95
	Ours	**66.51**	**38.36**	**62.59**	**35.44**	**62.59**
		(+8.10)	(+4.62)	(+7.64)	(+3.85)	(+7.64)

Question Reasoning. Question Reasoning. As shown in Table 2, we evaluate the results of our approach compared to RoG across four different scenarios involving the insertion of new questions, with differences in scores between the methods highlighted in green and red. When evaluated using only existing questions, our method's Hits@1, F1, and other metrics are lower than the baseline. This is due to two factors: (1) errors in the information extraction step that impacted downstream performance, and (2) certain answers in the new knowledge significantly aiding responses to questions in WebQSP, which cannot be accounted for during evaluation on WebQSP alone. Given these reasons, the test results on WebQSP are acceptable. From the last three sets of results in the table, it is clear that when we introduce new questions containing valuable knowledge, our method shows obvious superiority, especially when more new questions are added. When 900 new data points were added, our method's Hits@1 improved by 8.1% compared to RoG, demonstrating the effectiveness of our approach.

4.3   Case Study

Table 3. An example of our framework in WebQSP

Input	I only know that Sal Gibson is Keyshia Cole's father, and I know nothing else about her. I want to know who her parents are?
Information Extraction	Knowledge: Sal Gibson is Keyshia Cole's biological father Question: I want to know who her parents are?
Knowledge Updating	Transform: -> Sal Gibson is Keyshia Cole's biological father To: -> (Keyshia Cole, people.person.parents, Sal Gibson)
Path Planning	Keyshia Cole - people.person.parents - Daniel Hiram Gibson Jr. Keyshia Cole - people.person.parents - Francine Lons Keyshia Cole - people.person.parents - Leon Cole Keyshia Cole - people.person.children - Yvonne Cole Keyshia Cole - people.person.children - Sal Gibson Keyshia Cole - people.person.gender - Female
Answer	Sal Gibson, Francine Lons, Leon Cole

Table 3 presents a case running on WebQSP, the input, the results of information extraction, the method of normalization during knowledge updating, path planning, and the final summarized output. This is a linear process, and the results of each step are critical to producing the correct answer. It can be observed that the input is first divided into knowledge and questions, and then the LLM is used to convert the knowledge into triples within the knowledge graph. Path planning is a relationship-driven method for answer retrieval. In the final answer generation step, we follow RoG's approach, using the LLM to filter out unnecessary information based on the question and the output of the path planning to extract valuable and relevant information for the question.

5   Conclusions

This paper has presented a collaborative reasoning framework that combines large language models (LLMs) with knowledge graphs (KGs) to address challenges related to limited reasoning ability and lack of transparency when processing new knowledge. The framework integrates new knowledge by automatic prompt generation and dynamically updating KGs, improving the model's adaptability to novel domains. By utilizing the reasoning path generated by LLMs for planning retrieval, the framework not only enhances reasoning accuracy but also ensures that the reasoning process is transparent and verifiable. This approach offers a practical solution to the issue of knowledge limitation of LLMs in specialized fields. Therefore, it has great significance in a variety of practical applications.

References

1. Baek, J., Aji, A.F., Saffari, A.: Knowledge-augmented language model prompting for zero-shot knowledge graph question answering. arXiv preprint arXiv:2306.04136 (2023)
2. Bang, Y., et al.: A multitask, multilingual, multimodal evaluation of ChatGPT on reasoning, hallucination, and interactivity. arXiv preprint arXiv:2302.04023 (2023)
3. Betz, P., Meilicke, C., Stuckenschmidt, H.: Adversarial explanations for knowledge graph embeddings. In: IJCAI, vol. 2022, pp. 2820–2826 (2022)
4. Brown, T.B.: Language models are few-shot learners. arXiv preprint arXiv:2005.14165 (2020)
5. Chang, Y., et al.: A survey on evaluation of large language models. ACM Trans. Intell. Syst. Technol. **15**(3), 1–45 (2024)
6. Devlin, J.: BERT: pre-training of deep bidirectional transformers for language understanding. arXiv preprint arXiv:1810.04805 (2018)
7. Hamilton, W., Ying, Z., Leskovec, J.: Inductive representation learning on large graphs. In: Advances in Neural Information Processing Systems, vol. 30 (2017)
8. He, G., Lan, Y., Jiang, J., Zhao, W.X., Wen, J.R.: Improving multi-hop knowledge base question answering by learning intermediate supervision signals. In: Proceedings of the 14th ACM International Conference on Web Search and Data Mining, pp. 553–561 (2021)
9. He, R., et al.: On the effectiveness of adapter-based tuning for pretrained language model adaptation. arXiv preprint arXiv:2106.03164 (2021)
10. Hu, E.J., et al.: LoRA: low-rank adaptation of large language models. arXiv preprint arXiv:2106.09685 (2021)

11. Jiang, J., Zhou, K., Zhao, W.X., Wen, J.R.: UniKGQA: unified retrieval and reasoning for solving multi-hop question answering over knowledge graph. arXiv preprint arXiv:2212.00959 (2022)
12. Kaddour, J., Harris, J., Mozes, M., Bradley, H., Raileanu, R., McHardy, R.: Challenges and applications of large language models. arXiv preprint arXiv:2307.10169 (2023)
13. Lester, B., Al-Rfou, R., Constant, N.: The power of scale for parameter-efficient prompt tuning. arXiv preprint arXiv:2104.08691 (2021)
14. Li, F., Chen, M., Dong, R.: Multi-hop question answering with knowledge graph embedding in a similar semantic space. In: 2022 International Joint Conference on Neural Networks (IJCNN), pp. 01–07. IEEE (2022)
15. Li, S., et al.: Graph reasoning for question answering with triplet retrieval. arXiv preprint arXiv:2305.18742 (2023)
16. Li, T., Ma, X., Zhuang, A., Gu, Y., Su, Y., Chen, W.: Few-shot in-context learning for knowledge base question answering. arXiv preprint arXiv:2305.01750 (2023)
17. Li, X.L., Liang, P.: Prefix-tuning: optimizing continuous prompts for generation. arXiv preprint arXiv:2101.00190 (2021)
18. Liu, X., et al.: P-Tuning v2: prompt tuning can be comparable to fine-tuning universally across scales and tasks. arXiv preprint arXiv:2110.07602 (2021)
19. Luo, L., Li, Y.F., Haffari, G., Pan, S.: Reasoning on graphs: faithful and interpretable large language model reasoning. arXiv preprint arXiv:2310.01061 (2023)
20. Pan, S., Luo, L., Wang, Y., Chen, C., Wang, J., Wu, X.: Unifying large language models and knowledge graphs: a roadmap. IEEE Trans. Knowl. Data Eng. (2024)
21. Saxena, A., Tripathi, A., Talukdar, P.: Improving multi-hop question answering over knowledge graphs using knowledge base embeddings. In: Proceedings of the 58th Annual Meeting of the Association for Computational Linguistics, pp. 4498–4507 (2020)
22. Sun, H., Bedrax-Weiss, T., Cohen, W.W.: PullNet: open domain question answering with iterative retrieval on knowledge bases and text. arXiv preprint arXiv:1904.09537 (2019)
23. Sun, Y., Zhang, L., Cheng, G., Qu, Y.: SPARQA: skeleton-based semantic parsing for complex questions over knowledge bases. In: Proceedings of the AAAI Conference on Artificial Intelligence, vol. 34, pp. 8952–8959 (2020)
24. Sun, Z., Yang, H., Liu, K., Yin, Z., Li, Z., Xu, W.: Recent advances in LoRA: a comprehensive survey. ACM Trans. Sens. Netw. 18(4), 1–44 (2022)
25. Touvron, H., et al.: Llama 2: open foundation and fine-tuned chat models. arXiv preprint arXiv:2307.09288 (2023)
26. Vos, D., Döhmen, T., Schelter, S.: Towards parameter-efficient automation of data wrangling tasks with prefix-tuning. In: NeurIPS 2022 First Table Representation Workshop (2022)
27. Xu, Z., Jain, S., Kankanhalli, M.: Hallucination is inevitable: An innate limitation of large language models. arXiv preprint arXiv:2401.11817 (2024)
28. Yu, D., et al.: DecAF: joint decoding of answers and logical forms for question answering over knowledge bases. arXiv preprint arXiv:2210.00063 (2022)
29. Zhang, J., et al.: Subgraph retrieval enhanced model for multi-hop knowledge base question answering. arXiv preprint arXiv:2202.13296 (2022)
30. Zhang, R., et al.: Llama-adapter: efficient fine-tuning of language models with zero-init attention. arXiv preprint arXiv:2303.16199 (2023)
31. Zhou, Y., et al.: Large language models are human-level prompt engineers. arXiv preprint arXiv:2211.01910 (2022)

Zero-Shot Dense Retrieval Based on Query Expansion

Yaqi Wu[1] , Pengyu Chen[1] , Ziyu Ding[1] , and Anli Yan[2(✉)]

[1] Institute of Artificial Intelligence, Guangzhou University, Guangzhou, China
{winnerwu,dingziyu}@e.gzhu.edu.cn, cpyu66@gmail.com
[2] School of Artificial Intelligence, Guangzhou University, Guangzhou, China
anli_yan2021@163.com

Abstract. Dense retrieval is an effective information retrieval technique that utilizes semantic embedding similarity to retrieve documents. There are two typical kinds of dense retrieval methods, i.e., training-based methods and zero-shot methods. Training-based dense retrieval can achieve relatively high retrieval accuracy, but it relies on annotated datasets to train similarity retrieval models with high resource consumption. On the contrary, zero-shot dense retrieval does not require training-specific models, but its retrieval accuracy needs to be improved. To improve the accuracy of zero-shot dense retrieval, we propose a novel zero-shot dense retrieval based on query expansion (ZRQE). Specifically, it divides dense retrieval into two tasks: generating relevant documents using an instruction-following language model and dense retrieval by calculating vector similarity. The experimental results show that ZRQE achieves higher retrieval accuracy than other state-of-the-art dense retrieval methods.

Keywords: Zero-shot · Dense retrieval · Query expansion

1 Introduction

Dense retrieval is an efficient information retrieval technique that utilizes semantic embedding similarity to retrieve documents [14,35]. Dense retrieval has demonstrated its superiority in multiple application scenarios [6,36], especially in tasks of web search, question-answering systems, and fact verification. It can not only improve the accuracy of retrieval [12], but also significantly enhance the speed and efficiency of information acquisition [16].

At present, dense retrieval techniques are classified into training-based dense retrieval and zero-shot dense retrieval techniques depending on whether the neural network model needs to be trained or fine-tuned. Training-based dense retrieval technique relies on annotated datasets for training or fine-tuning to build and optimize models for vector similarity retrieval [24]. This technique improves the accuracy and relevance of retrieval tasks by training models to understand and recognize subtle differences and deep semantic connections

F. Zhang et al. (Eds.): AIS&P 2024, LNCS 15399, pp. 143–155, 2025.
https://doi.org/10.1007/978-981-96-1148-5_12

between texts [27]. Although the training-based dense retrieval technique provides higher semantic matching capabilities [15] and superior retrieval performance, the requirement of high resource consumption and massive data pose challenges in generalization ability for practical applications. On the contrary, the zero-shot dense retrieval technique does not require annotated dataset (i.e., relevance label dataset) training or fine-tuning of the model, but directly uses pre-trained models and employs text enhancement techniques to improve the accuracy of similarity retrieval [9]. While the zero-shot dense retrieval technique can operate without targeted training data, it achieves inferior performance in the applications of complex semantic understanding and fine-grained text comparison. Therefore, improving the ability of this technique to identify and match relevant texts accurately has become a hot research direction.

To address this issue, we propose a novel zero-shot retrieval based on the query expansion method, called ZRQE. It decomposes the process of dense retrieval into two tasks. The goal of the first task is to generate relevant document content based on a given query. For instance, an instruction-following language model is responsible for executing the generation task. The second task is performed by a contrastive encoder, which focuses on evaluating and determining the similarity between the generated vector documents and the existing vector documents in the database. Note that ZRQE does not conduct any new model training, whereas it relies on the pre-trained instruction-following language model and the contrastive encoder. This means that all processing and analysis processes are based on existing model frameworks, without the need for additional training processes to adapt to specific tasks or datasets. Specifically, ZRQE first inputs the query text into the search engine to obtain query-related web content. Then the instruction following language model uses the query and the obtained query-related web content to generate relevant documents according to the specific instruction description. Our instruction design enables the generative model to generate answers to queries based on query-related web content. For example, "$< web > \{webcontent\} < /web >$ Please write a passage to answer the question. You can selectively refer to the above web content. Question:{ query } Passage:", which is designed to guide the model to generate text that is closely related to the given query. We expect this generation process to effectively capture 'correlation' through a given query; Even if the generated document may not be entirely accurate and contain some factual errors, it should still resemble an actual document related to the query. Then, the generated document is input into a contrastive encoder, which is responsible for converting the document into a vectorized representation. Finally, these vectorized representations are compared for similarity with the document vectors already stored in the vector database. By this calculation process, ZRQE can identify and return the top k documents most relevant to the query.

Moreover, the reasons why we selected to expand queries through web content are given as follows. Firstly, relying solely on user queries for vector database retrieval can easily lead to retrieval bias, mainly because user queries may only be a short sentence, which has limited information and is difficult to fully reflect the user's real needs. Second, if the generative model is directly used to generate answers based on user queries, there is an obvious risk: the generative model may produce content unrelated to the problem or logically incoherent, known as "gibberish". Although such answers are generated, they cannot effectively solve the user's query problem. By introducing web content, we can provide richer contextual information for the model, thereby generating more accurate and relevant responses, effectively avoiding the above two problems. This method not only improves the accuracy of retrieval but also ensures the relevance and practicality of generated content. Our main contributions are summarized as follows.

- We propose a novel zero-shot retrieval based on the query expansion method, called ZRQE. This method does not require any model training.
- We propose to use instruction-following language models and query-related web content to extend the original query text, thus improving the accuracy of dense retrieval.
- We conduct a series of sufficient experiments to verify the effectiveness of the ZRQE method, including comparisons with other dense retrieval works in terms of retrieval accuracy.

RoadMap. Sect. 2 introduces related works, drawing comparisons and highlighting distinctions between our method and those previously established in the field. Section 3 presents the proposed zero-shot retrieval based on the query expansion method. Section 4 delves into the methods applied in our experiment, giving a comprehensive description of the data sets and metrics used. Empirical results are presented in Sect. 5. Section 6 summarizes this paper.

2   Related Works

Characterized by its technical attributes, dense retrieval techniques can be roughly divided into two types: training-based dense retrieval and zero-shot dense retrieval [35]. In this section, we analyze and summarize the advantages and disadvantages of each of the above two techniques. The overview diagram of training-based dense retrieval and zero-shot dense retrieval is shown in Fig. 1.

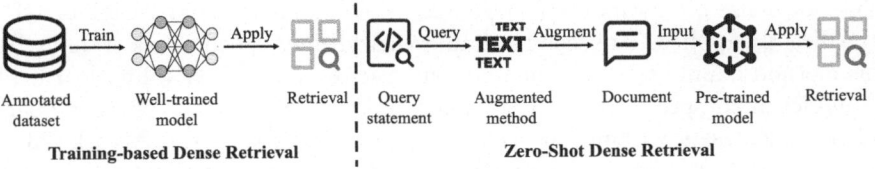

Fig. 1. Overview diagram of training-based dense retrieval and zero-shot dense retrieval

2.1   Training-Based Dense Retrieval

Training-based dense retrieval is an information retrieval technique that utilizes deep learning models to understand and compare the semantic information of queries and documents [4,34]. Unlike traditional keyword-based sparse retrieval methods, training-based dense retrieval generates dense vector representations of queries and documents through training models and then matches relevant documents by calculating the similarity between these vectors [10,17]. The key to training-based dense retrieval is effectively training the model to generate meaningful vector representations and quickly and accurately perform similarity calculations between vectors [22]. For example, Zhan et al. [32] proposed and analyzed a method to optimize dense search (DR) model training by using hard negative samples during training. It addresses the key challenge of training DR Models more effectively and efficiently, with a particular focus on the effects of different types of negative sampling and the dynamic nature of hard negatives during training. Yang et al. [30] proposed a new dense retrieval training method, particularly enhancing negative sampling techniques through a method called TriSampler. The core of the TriSampler revolves around the quasi-triangle principle, which intelligently guides the selection of negative samples within the triangular region. This method optimizes the training of the retrieval model by focusing on the negative term with the highest amount of information, thereby improving the performance of the model in distinguishing between related and unrelated documents.

Although training-based dense retrieval provides more accurate semantic matching capabilities, there are limitations as follows [31]: (1) Resource consumption. Dense retrieval relies on deep learning models to generate vector representations of documents and queries, which typically require significant computational resources, especially when using large models. (2) Training data dependency [33]. To effectively generate high-quality vector representations, dense retrieval models require massive training data. (3) Weak generalization ability [18]. Although training-based dense retrieval models perform well on training data, they may encounter generalization problems in unseen data or special situations, leading to a decrease in retrieval performance.

2.2   Zero-Shot Dense Retrieval

Zero-shot dense retrieval allows large language models (such as BERT or GPT) to be directly applied to unseen retrieval tasks without specific task training [19]. This method mainly relies on the rich semantic representation ability learned by the model in the pre-training stage, allowing it to directly process new queries and document sets without the need for fine-tuning for each new task [23]. In zero-shot dense retrieval, the model generates a dense vector representation of the query and document and then retrieves the relevant document by calculating the similarity between these vectors [29]. The key advantage of this method is its high versatility and flexibility, as the same pre-trained model can cope with many different types of retrieval tasks without the requirement of retraining for each one [9]. This technique is particularly suitable for resource-constrained or data-scarce environments, since it reduces the reliance on massive annotated data, while also improving the efficiency of model deployment [9]. Zero-shot dense retrieval has promoted the development of natural language processing, especially in improving the scalability and adaptability of information retrieval [25]. For example, Gao et al. [5] proposed a zero-shot dense retrieval method, i.e., HyDE. HyDE does not require any correlation labels and performs high generalizability across different tasks and languages. The method first utilizes an instruction tracking language model, such as InstructGPT, to generate a hypothetical document based on a given query. The hypothetical document is then encoded as an embedded vector using an unsupervised contrastive learning encoder such as Contriever. Fang et al. [3] proposed a robust zero-shot dense retrieval method called RMSC (Robust Multi-Supervision Combining). RMSC solves the domain adaptation challenge in dense retrieval models by clearly distinguishing between domain data and supervised signals during the training process. Specifically, RMSC improves performance by combining weakly supervised data from the target domain with labeled data from the source domain, using these soft tokens to store domain-specific and supervised knowledge, thereby avoiding overfitting and improving generalization ability across different domains.

Although zero-shot dense retrieval has been widely studied as an advanced information retrieval technique, existing zero-shot dense retrieval methods still face some challenges, especially in terms of retrieval accuracy [21]. Since this method relies on the general semantic representation learned by the model in the pre-training stage, without optimization for specific retrieval tasks [26], when dealing with highly specialized queries or domain-specific documents, its performance may not be as good as models trained for specific tasks [20]. In this paper, we proposed a novel zero-shot retrieval based on the query expansion method, called ZRQE. This method does not require any model training.

3   Methodology

3.1   Preliminaries

Zero-Shot Dense Retrieval. The dense retrieval mainly achieves effective information retrieval by calculating the inner vector product similarity between queries and documents. This method relies on transforming queries and documents into vectors in a high-dimensional space and then evaluating their similarity by measuring the dot product between them. Given a query text q and a document d, it processes them separately using two encoder functions $encode_q$ and $encode_d$. We map these two text entities to an N-dimensional vector space to obtain vectors v_q and v_d. The equation for calculating the similarity between queries and documents is given in Eq. (1).

$$Sim(q, d) = <encode_q(q), encode_d(d)> = <v_q, v_d> \tag{1}$$

In the scenario of zero-shot retrieval, we operate on m sets of query sets Q_1, $Q_2,...,Q_m$ and their corresponding document sets $D_1, D_2,...,D_m$. For each set Q_i, we use q_{ij} to identify the j-th query. Our task is to build $encode_q$ and $encode_d$, which can map queries and documents to the same embedding space. During this process, we do not need to access any specific query set Q_i, document set D_i, or known correlation labels r_{ij} between them. This setup requires our model to rely on pre-trained knowledge and semantic understanding to automatically identify the relationship between queries and documents in the absence of clear relevance feedback.

Instruction-Following Language Model. The instruction-following language model is a specialized language model designed to parse and execute specific textual instructions. The training of this model aims to learn how to understand and respond to specific task descriptions given by users, such as paraphrasing text, answering questions, or generating content that meets specific requirements. Formally, this model can be represented as a function f that takes an instruction I and a possible context C as inputs and outputs a response A. The function is defined as $A = f(I, C)$. This type of model is typically pre-trained on large-scale datasets, learning to extract intent from multiple instructions and generate appropriate responses, optimizing its accuracy and flexibility in executing complex tasks.

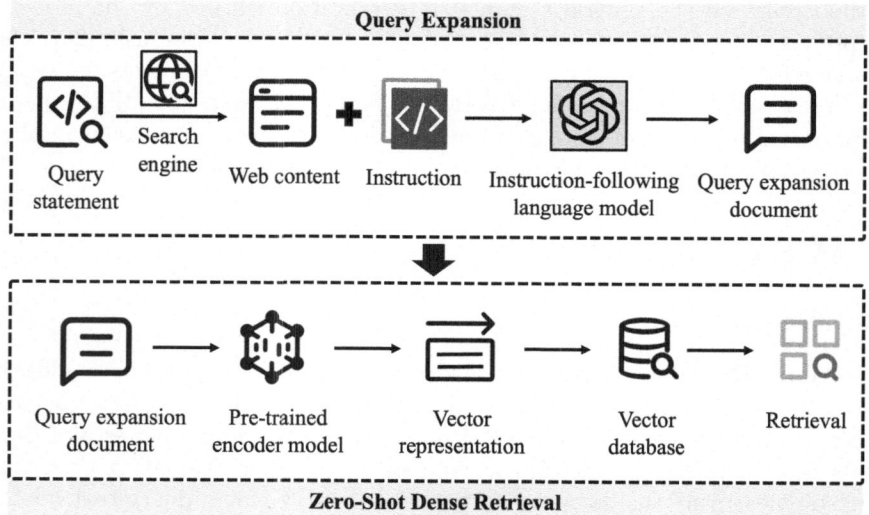

Fig. 2. Schematic depiction of the analysis steps in ZRQE

3.2   ZRQE

ZRQE is mainly composed of two core components, namely query expansion and zero-shot dense retrieval, as shown in Fig. 2. The function of query expansion is to extend the original query, to generate richer and more valuable knowledge by adding relevant contextual information, thereby improving the relevance and accuracy of retrieval. Zero-shot dense retrieval is responsible for dense retrieval, which receives queries extended by query expansion, converts these queries into vectors, and then performs similarity retrieval in the vector database.

Query Expansion. The query text is first input into the search engine to obtain query-related web content, and then the query text and query-related web content are input into the instruction-following language model according to the given instruction format, which is specially used to generate relevant documents based on the input. In this paper, the instruction-following language model we used is GPT-4o Mini, which is a generative model capable of understanding and executing complex language tasks. The specific instructions we provide for the GPT-4o Mini are: "$< web > \{webcontent\} < /web >$ Please write a passage to answer the question. You can selectively refer to the above web content. Question:{query} Passage:". This process involves an in-depth analysis of the topics involved in the query and utilizing the provided web content as reference information to ensure the generation of documents relevant to the query, even if there are some errors.

Given a query text q, w represents the web content related to the query obtained via the search engine, and I represents the instruction associated with it. The instruction-following language model is represented as M_{GPT}. We can describe the behavior of the model as a function f, as shown in the equation Eq.(2). Among them, D is the document generated by the model. This function f is responsible for processing query text q, query-related web content w, and instructions I, utilizing the language generation capability of GPT-4o Mini to generate documents.

$$D = f(M_{GPT}, q, w, I) \tag{2}$$

Zero-Shot Dense Retrieval. Once we obtain the generated document D_i, which is the expanded query text, we encode the generated document. We can use encoder $encode_q$ to encode the generated document.

$$v_i = encode_q(D_i) \tag{3}$$

Subsequently, we retrieve the most similar content in the vector database $U = U_1, U_2, U_3, ...U_L$ based on V_i. L is the number of documents in a vector database. As mentioned above, we measure similarity by calculating the inner product of vectors. The equation for calculating the vector inner product is shown in Eq.(4).

$$Sim(v_{ij}, u_{lr}) = \sum_{k=1}^{n} v_{ij}^k u_{lr}^k \quad u_{lr} \in U_l \quad v_{ij} \in V_i \tag{4}$$

where v_{ij} is the j-th sentence in the i-th query extension document, u_{lr} is the r-th sentence in the l-th document in the vector database, and n is the vector dimension.

4   Experimental Design

4.1   Implementation

We implemented ZRQE using GPT-4o Mini, a compact variant of OpenAI's GPT-4o model, alongside the Contriever models [7]. For open-ended text generation, we sampled outputs from GPT-4o Mini, using the default temperature setting of 0.7. The Contriever model was employed for retrieval tasks, and all retrieval experiments were conducted using the Pyserini toolkit [13].

4.2   Dataset

We utilize TREC DL19 [2] and DL20 datasets for dense retrieval task experiments; they are based on the MS-MARCO dataset [1]. We also used three low-resource datasets, i.e., Scifact, Trec-Covid, and Trec-News for experiments, all of which are from the BEIR dataset [21]. We use different directives for different datasets, which have similar structures but with different quantifiers to control the exact form of the generated documents.

4.3   Metrics

In our experiments, we evaluate the performance of our dense retrieval method using three widely recognized metrics in information retrieval: Mean Average Precision (**MAP**), Recall at rank 1000 (**Recall@1k**), and Normalized Discounted Cumulative Gain at rank 10 (**nDCG@10**). These metrics are selected to capture various aspects of the retrieval effectiveness, including precision, relevance ranking, and recall. MAP measures the precision of the retrieval system across all queries, focusing on the relevance of retrieved documents. nDCG@10 is used to assess the ranking quality of the top 10 retrieved documents. It considers not only the relevance of the documents but also their positions in the ranked list. Recall@1k evaluates the ability of the model to retrieve as many relevant documents as possible within the top 1000 results.

4.4   Compared Baselines

Contriever serves as our main baseline, which is trained using unsupervised contrastive learning. Our method ZRQE retriever shares the same embedding space with it. The only difference is in the way the query vector is constructed. These comparisons make it easy to examine the effect of ZRQE. In addition, we also compare with the classic heuristic-based lexical retriever BM25 [28], which does not require supervision.

Several methods for fine-tuning using massive relevance data are also included as references. We consider DPR [8] and ANCE [28], which are trained on supervised data such as NaturalQuestions [11] or MS MARCO. We also include the state-of-the-art transfer learning models: Contriever fine-tuned on MS-MARCO, denoted Contriever[FT]. This model has been trained using the state-of-the-art retrieval model training pipeline that involves second-stage of retrieval-specific pre-training; it should be considered an empirical upper bound.

Table 1. Results for dense retrieval on TREC DL19/20. Best performance w/o supervision and overall system(s) are marked **bold**. DPR, ANCE, and Contriever[FT] are in-domain supervised models that are finetuned on MS MARCO training data

Model	TREC DL19			TREC DL20		
	MAP	nDCG@10	Recall@1k	MAP	nDCG@10	Recall@1k
w/o supervision						
BM25	30.1	50.6	75.0	28.6	48.0	78.6
Contriever	24.0	44.5	74.6	24.0	42.1	75.4
ZRQE	**41.9**	**61.5**	**88.3**	**39.2**	**59.4**	**86.1**
w/ supervision						
DPR	36.5	62.2	76.9	41.8	**65.3**	81.4
ANCE	37.1	**64.5**	75.5	40.8	64.6	77.6
Contriever[FT]	41.7	62.1	83.6	**43.6**	63.2	85.8

Table 2. Low resource tasks from BEIR. Best performance w/o supervision and overall system(s) are marked in **bold**

	Scifact	Trec-Covid	Trec-News
w/o supervision			
BM25	67.9	**59.5**	39.5
Contriever	64.9	27.3	34.8
ZRQE	**69.2**	59.1	**43.8**
w/ supervision			
DPR	31.8	33.2	16.1
ANCE	50.7	**65.4**	38.2
Contriever[FT]	67.7	59.6	42.8

5   Empirical Results

5.1   Dense Retrieval

We conduct dense retrieval experiments on the TREC DL19 and TREC DL20 datasets. In Table 1, we present the retrieval results for both datasets, demonstrating that our method, i.e., ZRQE, yields significant improvements across the board for Contriever. These improvements are evident in both precision-oriented and recall metrics. Although the unsupervised Contriever model may underperform compared to the traditional BM25 approach, our method ZRQE significantly outperforms BM25 by a wide margin.

From the experimental results, it is clear that ZRQE remains competitive even when compared with the models fine-tuned with massive amounts of relevant data. It is worth noting note that TREC DL19/20 are the search tasks based on MS-MARCO, where all fine-tuned models benefit from extensive supervision. On TREC DL19, ZRQE achieves comparable MAP and nDCG@10 scores to Contriever^{FT}, while significantly outperforming Contriever^{FT} in recall@1k. On TREC DL20, ZRQE performs approximately 10% lower than Contriever^{FT} in MAP and nDCG@10 but outperforms it on recall@1k. In addition, although the ANCE model achieves higher nDCG@10 scores than ZRQE, it suffers from lower recall. This suggests that it may be biased towards a subset of queries and/or related documents, resulting in poor overall ranking performance.

5.2   Low Resource Retrieval

In Table 2, we show the retrieval results on three low-resource retrieval task datasets from BEIR. Experimental results show that ZRQE also brings significant improvements to Contriever on low-resource retrieval tasks. ZRQE only performs worse than BM25 on one dataset, Trec-Covid, but our method brings more than 50% performance improvement compared to Contriever.

In addition, experimental results also show that our method ZRQE has obvious advantages over the fine-tuned models. ZRQE generally shows better performance than ANCE and DPR, even though they are fine-tuned on MS-MARCO.

Table 3. nDCG@10 on TREC DL19/20. Effect of changing different LLMs

Model	TREC DL19	TREC DL20
Contriever	44.5	42.1
ZRQE		
w/ Cohere (52b)	54.0	53.9
w/ GPT-4o Mini (175b)	61.5	59.4
w/ GPT-4 (-)	**63.2**	**60.4**

5.3 Effect of Different Generative Models

In Table 3, we show ZRQE using other large language models. Specifically, we consider a 52 billion Cohere model and a stronger model GPT4, which is currently the most powerful language model. Empirically, the texts generated by smaller language models tend to be shorter and contain more factual errors. Therefore, according to the experimental results, it is clear that larger and better language models often achieve better performance improvements.

6 Conclusion

This paper has presented ZRQE, which is a novel zero-shot dense retrieval based on the query expansion method. ZRQE relies on a pre-trained instruction-following language model and contrastive encoder to decompose dense retrieval into generating relevant documents based on query extensions and calculating similarity with vector database documents based on the vector representations of the generated documents, thereby effectively processing queries without training a new model. The experimental results demonstrate that our proposed ZRQE method outperforms other state-of-the-art dense retrieval methods in the aspect of retrieval accuracy. This advantage indicates that ZRQE can more effectively understand and process complex queries, providing more accurate and relevant search results.

References

1. Bajaj, P., et al.: MS MARCO: a human generated machine reading comprehension dataset. arXiv preprint arXiv:1611.09268 (2016)
2. Craswell, N., Mitra, B., Yilmaz, E., Campos, D., Voorhees, E.M.: Overview of the TREC 2019 deep learning track. arXiv preprint arXiv:2003.07820 (2020)
3. Fang, Y., Ai, Q., Zhan, J., Liu, Y., Wu, X., Cao, Z.: Combining multiple supervision for robust zero-shot dense retrieval. In: Proceedings of the AAAI Conference on Artificial Intelligence, vol. 38, pp. 17994–18002 (2024)
4. Gao, L., Callan, J.: Condenser: a pre-training architecture for dense retrieval. arXiv preprint arXiv:2104.08253 (2021)
5. Gao, L., Ma, X., Lin, J., Callan, J.: Precise zero-shot dense retrieval without relevance labels. In: Rogers, A., Boyd-Graber, J.L., Okazaki, N. (eds.) Proceedings of the 61st Annual Meeting of the Association for Computational Linguistics (Volume 1: Long Papers), ACL 2023, Toronto, Canada, July 9-14, 2023, pp. 1762–1777. Association for Computational Linguistics (2023)
6. Huang, Z., Zeng, H., Zamani, H., Allan, J.: Soft prompt decoding for multilingual dense retrieval. In: Proceedings of the 46th International ACM SIGIR Conference on Research and Development in Information Retrieval, pp. 1208–1218 (2023)
7. Izacard, G., et al.: Towards unsupervised dense information retrieval with contrastive learning. arXiv preprint arXiv:2112.09118 (2021)
8. Karpukhin, V., et al.: Dense passage retrieval for open-domain question answering. arXiv preprint arXiv:2004.04906 (2020)

9. Khramtsova, E., Zhuang, S., Baktashmotlagh, M., Wang, X., Zuccon, G.: Selecting which dense retriever to use for zero-shot search. In: Proceedings of the Annual International ACM SIGIR Conference on Research and Development in Information Retrieval in the Asia Pacific Region, pp. 223–233 (2023)

10. Kong, W., et al.: Multi-aspect dense retrieval. In: Proceedings of the 28th ACM SIGKDD Conference on Knowledge Discovery and Data Mining, pp. 3178–3186 (2022)

11. Kwiatkowski, T., et al.: Natural questions: a benchmark for question answering research. Trans. Assoc. Comput. Linguis. **7**, 453–466 (2019)

12. Li, C., Liu, Z., Xiao, S., Shao, Y., Lian, D.: Llama2Vec: unsupervised adaptation of large language models for dense retrieval. In: Proceedings of the 62nd Annual Meeting of the Association for Computational Linguistics (Volume 1: Long Papers), pp. 3490–3500 (2024)

13. Lin, J., Ma, X., Lin, S.C., Yang, J.H., Pradeep, R., Nogueira, R.: Pyserini: a Python toolkit for reproducible information retrieval research with sparse and dense representations. In: Proceedings of the 44th International ACM SIGIR Conference on Research and Development in Information Retrieval, pp. 2356–2362 (2021)

14. Lin, S., et al.: How to train your dragon: diverse augmentation towards generalizable dense retrieval. In: Bouamor, H., Pino, J., Bali, K. (eds.) Findings of the Association for Computational Linguistics: EMNLP 2023, Singapore, December 6-10, 2023, pp. 6385–6400. Association for Computational Linguistics (2023)

15. Lin, S.C., Lin, J.: A dense representation framework for lexical and semantic matching. ACM Trans. Info. Syst. **41**(4), 1–29 (2023)

16. Lin, Z., et al.: PROD: progressive distillation for dense retrieval. In: Proceedings of the ACM Web Conference 2023, pp. 3299–3308 (2023)

17. Ma, X., Teofili, T., Lin, J.: Anserini gets dense retrieval: integration of Lucene's HNSW indexes. In: Proceedings of the 32nd ACM International Conference on Information and Knowledge Management, pp. 5366–5370 (2023)

18. Prakash, P., Killingback, J., Zamani, H.: Learning robust dense retrieval models from incomplete relevance labels. In: Proceedings of the 44th International ACM SIGIR Conference on Research and Development in Information Retrieval, pp. 1728–1732 (2021)

19. Ren, R., et al.: A thorough examination on zero-shot dense retrieval. In: Bouamor, H., Pino, J., Bali, K. (eds.) Findings of the Association for Computational Linguistics: EMNLP 2023, Singapore, December 6-10, 2023, pp. 15783–15796. Association for Computational Linguistics (2023)

20. Shin, G., Xie, W., Albanie, S.: ReCo: retrieve and co-segment for zero-shot transfer. Adv. Neural. Inf. Process. Syst. **35**, 33754–33767 (2022)

21. Thakur, N., Reimers, N., Rücklé, A., Srivastava, A., Gurevych, I.: BEIR: a heterogenous benchmark for zero-shot evaluation of information retrieval models. arXiv preprint arXiv:2104.08663 (2021)

22. Tonellotto, N., Macdonald, C.: Query embedding pruning for dense retrieval. In: Proceedings of the 30th ACM International Conference on Information and Knowledge Management, pp. 3453–3457 (2021)

23. Wang, X., Macdonald, C., Ounis, I.: Improving zero-shot retrieval using dense external expansion. Info. Process. Manag. **59**(5), 103026 (2022)

24. Wang, X., Macdonald, C., Tonellotto, N., Ounis, I.: ColBERT-PRF: semantic pseudo-relevance feedback for dense passage and document retrieval. ACM Trans. Web **17**(1), 1–39 (2023)

25. Wu, L., Petroni, F., Josifoski, M., Riedel, S., Zettlemoyer, L.: Scalable zero-shot entity linking with dense entity retrieval. In: Webber, B., Cohn, T., He, Y., Liu, Y. (eds.) Proceedings of the 2020 Conference on Empirical Methods in Natural Language Processing, EMNLP 2020, Online, November 16-20, 2020, pp. 6397–6407. Association for Computational Linguistics (2020)

26. Wu, T., Bai, X., Guo, W., Liu, W., Li, S., Yang, Y.: Modeling fine-grained information via knowledge-aware hierarchical graph for zero-shot entity retrieval. In: Proceedings of the Sixteenth ACM International Conference on Web Search and Data Mining, pp. 1021–1029 (2023)

27. Wu, X., Ma, G., Lin, M., Lin, Z., Wang, Z., Hu, S.: Contextual masked auto-encoder for dense passage retrieval. In: Proceedings of the AAAI Conference on Artificial Intelligence, vol. 37, pp. 4738–4746 (2023)

28. Xiong, L., et al.: Approximate nearest neighbor negative contrastive learning for dense text retrieval. arXiv preprint arXiv:2007.00808 (2020)

29. Xu, C., Guo, D., Duan, N., McAuley, J.J.: LaPraDoR: unsupervised pretrained dense retriever for zero-shot text retrieval. In: Muresan, S., Nakov, P., Villavicencio, A. (eds.) Findings of the Association for Computational Linguistics: ACL 2022, Dublin, Ireland, May 22-27, 2022, pp. 3557–3569. Association for Computational Linguistics (2022)

30. Yang, Z., Shao, Z., Dong, Y., Tang, J.: TriSampler: a better negative sampling principle for dense retrieval. In: Proceedings of the AAAI Conference on Artificial Intelligence, vol. 38, pp. 9269–9277 (2024)

31. Yu, H., Xiong, C., Callan, J.: Improving query representations for dense retrieval with pseudo relevance feedback. In: Proceedings of the 30th ACM International Conference on Information and Knowledge Management, pp. 3592–3596 (2021)

32. Zhan, J., Mao, J., Liu, Y., Guo, J., Zhang, M., Ma, S.: Optimizing dense retrieval model training with hard negatives. In: Proceedings of the 44th International ACM SIGIR Conference on Research and Development in Information Retrieval, pp. 1503–1512 (2021)

33. Zhan, J., Mao, J., Liu, Y., Guo, J., Zhang, M., Ma, S.: Learning discrete representations via constrained clustering for effective and efficient dense retrieval. In: Proceedings of the Fifteenth ACM International Conference on Web Search and Data Mining, pp. 1328–1336 (2022)

34. Zhang, X., Ogueji, K., Ma, X., Lin, J.: Toward best practices for training multilingual dense retrieval models. ACM Trans. Info. Syst. **42**(2), 1–33 (2023)

35. Zhao, W.X., Liu, J., Ren, R., Wen, J.R.: Dense text retrieval based on pretrained language models: a survey. ACM Trans. Info. Syst. **42**(4), 1–60 (2024)

36. Zhou, T., et al.: MARVEL: unlocking the multi-modal capability of dense retrieval via visual module plugin. In: Proceedings of the 62nd Annual Meeting of the Association for Computational Linguistics (Volume 1: Long Papers), pp. 14608–14624 (2024)

Lightweight Attention-CycleGAN for Nighttime-Daytime Image Transformation

Junhao Huang[1], Xiangjun Xiao[1], Haojun Zhou[1], Affan Yasin[2],
and Zhili Zhou[3]($\boxtimes$)

[1] School of Computer Science and Cyber Engineering, Guangzhou University,
Guangzhou, China
[2] Faculty of Science and Technology, Universitas Airlangga, Surabaya, Indonesia
[3] School of Artificial Intelligence, Guangzhou University, Guangzhou, China
zhou_zhili@163.com

Abstract. With the rapid development of deep learning in the field of computer vision, the performance of core vision tasks such as image recognition has achieved significant improvement. In nighttime environment, due to the low-light condition and reduced visibility, cross-domain transformation of nighttime images based on Generative Adversarial Network (GAN) model can effectively improve the accuracy of nighttime recognition models. However, the existing GAN models are difficult to be effectively deployed on resource-constrained devices due to the requirement of high storage space and computational resource. To this end, this paper proposes a shared attention network based on the attention mechanism with the CycleGAN structure, and designs an online knowledge distillation method to compress and optimize the model, so as to obtain a lightweight model to achieve the nighttime-daytime cross-domain image transformation. Experimental results demonstrate that the proposed model achieves the state-of-the-art performance in the task of Nighttime-Daytime Image Transformation. This is of great significance for edge devices to solve the problem of recognition at night.

Keywords: Image-to-Image Transformation · CycleGAN · Attention Guided · Knowledge Distillation

1 Introduction

As a critical subfield of machine learning, deep learning has emerged as a research hotspot in recent years. Its fundamental principle involves modeling complex information via multilayer neural network architectures, enabling the effective extraction and representation of high-level features from data. With the great advancements in deep learning, notable progress has been achieved in computer vision. Image recognition, a core challenge in this domain, aims to automatically detect, classify, and interpret objects or scenes within images. As a widely

F. Zhang et al. (Eds.): AIS&P 2024, LNCS 15399, pp. 156–166, 2025.
https://doi.org/10.1007/978-981-96-1148-5_13

adopted image analysis technique, image recognition plays an important role in various applications, including facial recognition, autonomous driving, and medical image analysis. The task typically involves mapping input images to corresponding categorical labels, thereby facilitating the identification and classification of objects within the image. Deep learning-based image recognition techniques, particularly those Convolutional Neural Networks (CNNs), are usually trained on large-scale image datasets, enabling enhanced performance and accuracy in recognition.

However, in low-illumination environments such as nighttime scenes, the accuracy of image recognition is significantly reduced [1]. That is because the structural, textural, and color features of objects may deteriorate or vanish in low-light conditions due to insufficient illumination or noise, leading to a substantial decline in recognition performance [2]. To address this issue, to achieve accurate image recognition in low-illumination environments, it is reasonable to transform nighttime images to the their daytime versions via cross-domain image translation at first. Although Generative Adversarial Networks (GANs) have made success in image synthesis [3] and image-to-image transformation [4], they generally require huge computation resource, which becomes a key bottleneck when these networks are deployed on resource-constrained cell phones or other lightweight IoT devices [5]. Therefore, in this study, we design a lightweight attention-CycleGAN for nighttime-daytime image transformation, optimize the structure to further improve the quality of the transformed images, and propose a GAN-oriented online distillation strategy to obtain a compression model, which is important for edge devices to achieve desirable performance of image detection and recognition in low illumination scenarios.

2   Related Work

2.1   Generative Adversarial Networks (GANs)

Generative Adversarial Networks (GANs) [6] are powerful generative models that have achieved impressive performances in different computer vision tasks such as image/video generation [7]. In order to generate meaningful images that satisfy the user's needs, a conditional GAN (CGAN) [8] injects additional information to guide the image generation process, which can be discrete labels [9], object keypoints [10], human skeleton [11], and semantic maps [12]. Image-to -Image Translation model uses CNNs to learn translation functions.Pix2pix [4] is a conditional framework that uses CGAN to learn mapping functions between input to output images. Wang et al. proposed Pix2pixHD [13] for high resolution photo-level image generation. Unpaired Image to Image Translation. To overcome this limitation, unpaired image to image translation task has been proposed. In this task, these methods learn mapping functions without the requirement of paired training data. Specifically, CycleGAN [14] learns a mapping function between two image domains instead of paired images. In addition to CycleGAN, many other GAN variants [15] have been proposed to solve the cross-domain problem.

However, these models cannot generate the most discriminative semantic parts of the image during the translation phase.

2.2 Attention-Guided Image-to-Image Translation

In order to address the above limitations, some works employ attention mechanisms to realize image translation. Attention mechanisms have been successfully applied to many applications in computer vision, such as depth estimation [16], to help models focus on relevant parts of the input. For example, Liang et al. proposed ContrastGAN [17], which uses objects to mask annotations from each dataset as additional input data. In addition, the basis of the so-called "self-attention" mechanism is introduced [18], which was later used as a major component of the Transformer Architecture [19] and was subsequently adapted to perform computer vision tasks with good performance in classification and target detection [20]. The Efficient Attention mechanism [21] proposes a mathematically equivalent self-attention mechanism by utilizing the associative property of matrix multiplication. Unlike self-attention, this method interprets each channel of the key as a global attention map, which does not correspond to a specific location but to a semantic aspect of the input image.

2.3 Knowledge Distillation

Knowledge distillation has become a conventional method for model compression since FitNet was proposed in 2014 [22], and is especially widely used in the field of generative adversarial network (GAN) compression. The core idea is to transfer knowledge from a larger model to a smaller model to improve the performance of the smaller one. Researchers have made a lot of efforts to mine a variety of important knowledge cues, such as output logits [23], intermediate feature maps [24], and instance relations [25]. The research in this paper is mainly inspired by the intermediate feature graph distillation method, which has been widely used to effectively guide the training of student networks. Compared with other knowledge cues, feature maps usually contain richer information and provide more detailed guidance for student networks. However, traditional knowledge distillation methods usually require pixel-by-pixel matching between the feature maps of the teacher and student networks, which is not ideal for GAN compression tasks, since the goal of GAN is to generate perceptually similar images.

In this paper, we utilize an efficient attention module and its unique interpretation of key passages as an attention graph to implement an attention-based architecture and attentional mechanism for the I2I translation task with an online distillation scheme aimed at learning effective student models from the complementary structures of the teacher generator and the different knowledge layers to achieve model lightness.

3 Methodology

In this section, we first introduce the proposed attention sharing strategy in Sect. 3.1 to construct novel generators; then Sect. 3.2 introduces online GAN distillation, in which the student generators are not constrained by the discriminator and try to learn concepts directly from the teacher model.

3.1 Attention Sharing

In the context of a generic image-to-image (I2I) translation problem, we consider two image domains, i.e., X and Y, each containing data instances $x \in X$ and $y \in Y$. respectively. The objective of this work is to learn two mapping functions represented by the generator G and F: $G : x \rightarrow y$ and $F : y \rightarrow x$, such that the distributions $G(X)$ and $F(Y)$ are indistinguishable from Y and X, respectively.

In an image-to-image translation task, the source image and the ideal translated image are structurally similar, and the attention graphs of the source and target domains will exhibit a high degree of similarity. Therefore, similarity in translation of scene elements can be achieved by sharing the attention graph from the source nighttime domain to the target daytime domain. An effective attention mechanism treats each key $K \in \mathbb{R}^{n \times d_k}$ as an independent attention graph, rather than realizing point-to-point attention in the semantics of the image. Given the high correlation between the content of the input and output images, the attention from the previous layers of the generator architecture can be reused in the corresponding subsequent layers. Since the semantic content of the input and output images of the generator G are highly correlated, it can be inferred that for the i convolutional layer of the encoder and the $i - L$ convolutional layers of the corresponding decoder, the Key generated by its shared attention algorithm can be reused. The original attention maps computed from the nighttime images can create a more realistic output when reconstructing the translated images. The shared attention weights are calculated by

$$O = \sigma_{row}\left(\frac{Q}{\sqrt{d_k}}\right)\sigma_{col}\left(\frac{K}{\sqrt{d_k}}\right)^T \tag{1}$$

where σ_{row} and σ_{col} is the application of the softmax function along each row or column of the matrix, d_k the dimensionality of the key vectors, Q the query vector, and K the key vector, respectively.

In the shared attention mechanism, long-term dependencies are captured through skip connections between key vectors. A refinement network is inserted between these skip connections to refine the attention maps of the original domain. To this end, we employ a deep convolutional network equipped with 3×3 kernels, which partitions the input channels into several groups, each of which is processed by distinct convolutional kernels. In this architecture, d_k groups are used to ensure that each channel originating from the Keys is convolved through its corresponding kernel, thereby preventing the mixing of different attention maps during the refinement process. This attention sharing mechanism is illustrated in Fig. 1.

Given that the network is more likely to learn dependencies in local neighbourhoods, it makes more sense to want to rely more on these local dependencies and wait for remote dependencies to become available during the initial training step. To achieve this, a learnable scale parameter is introduced to multiply the output of the attention and then add the input features. Where γ is initialised to zero to initially focus on the easier task of generating good features based on local neighbourhoods with convolution, and then assigning higher weights to the remote dependencies learned from the more meaningful features. In this way, the final output of the introduced shared attention algorithm is given by the following equation:

$$y_i = \gamma O_i + x_i \tag{2}$$

Analogous to the increasing number of channels with layer depth in deep convolutional neural networks, the number of attention maps d_k in this implementation is set to half the number of input channels per layer. Considering the significance of the initial attention block, it learns dependencies directly from the original three-channel input, and for the d_k dimension, the number of attention maps in the initial block is fixed at 8. Figure 2 displays this shared generator architecture.

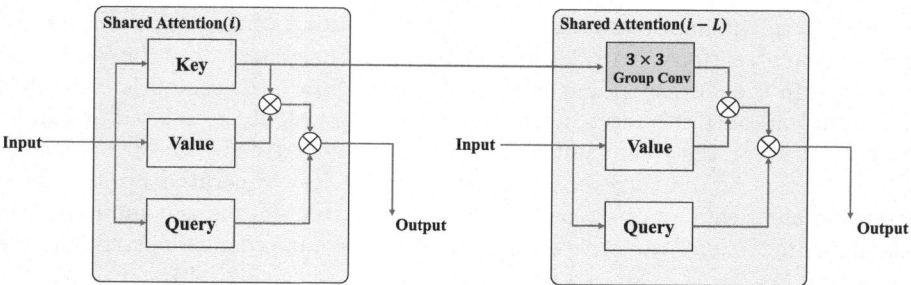

Fig. 1. Attention sharing Mechanism includes the Key, Value, and Query implementations. An effective attention mechanism interprets the key results as an attention graph shared with the corresponding i and $i - L$ layers. The attention graph is refined by 3×3 grouped convolutions to refine the attention graph while ensuring that there is no mixing of information between different channels

3.2   Online GAN Distillation

In this work, we employ an online distillation algorithm to compress the model, utilizing the generator G as the teacher generator G_T. Knowledge is transferred from the teacher network to the student generator G_S, ensuring that critical knowledge is effectively migrated. This process facilitates the student generator's understanding and learning of the task, ultimately training a lightweight student generator model. By using the student generator as the image generator, we address the issue of high complexity in the original generator. This online GAN knowledge distillation process is shown in the Fig. 3.

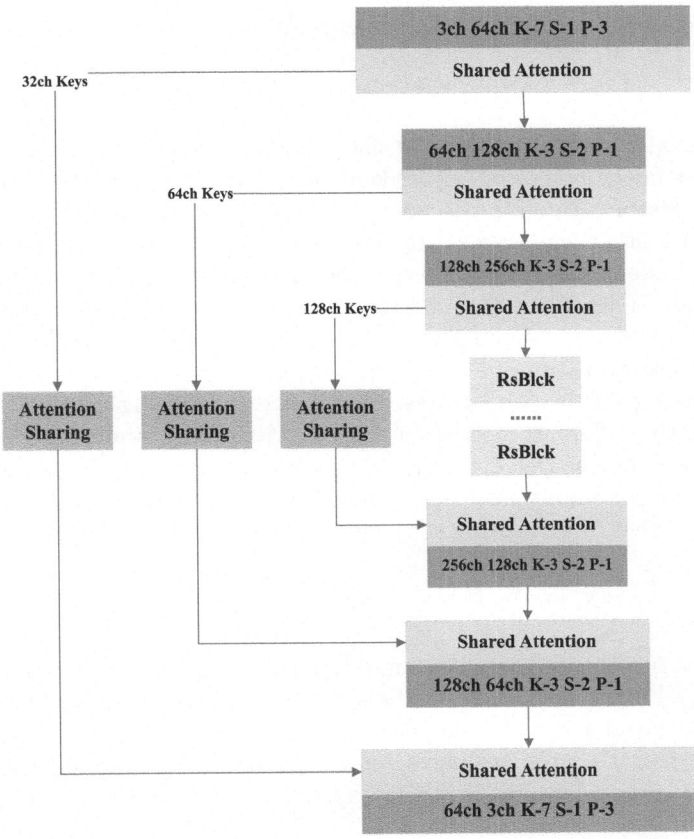

Fig. 2. The proposed shared attention generator architecture. The architecture introduces efficient attention modules on the encoder and decoder to obtain the best performance as shown in Fig. 1. The introduced attention module utilizes the attention sharing mechanism shown in 1

The objective of G_T is to learn the mapping function that transforms images from the nighttime domain X to the daytime domain Y. We train the generator G_T to map X_i to Y_i, utilizing a discriminator D to distinguish between the images generated by G_T and the real images. The teacher generator G_T generates feature maps of the nighttime domain image, denoted as F_{gt}, via an encoder. These feature maps are then decoded into daytime domain images $G_{T(x)}$ via a decoder. To ensure that the daytime domain images generated by $G_{T(x)}$ closely resemble the real daytime domain images y, we minimize the following reconstruction loss:

$$\mathcal{L}_{Recon}(G_T) = \mathbb{E}_{x,y}\left[\| \, y - G_T(x) \, \|_1\right] \tag{3}$$

where y is the real daytime domain image and $G_{T(x)}$ is the daytime domain image generated by the generator.

Therefore, the loss function and the optimization objective are defined as:

$$G_T^* = \arg\min_{G_T}\max_{D} \mathcal{L}_{GAN}\left(G_T, D\right) + \mathcal{L}_{Recon}\left(G_T\right) \tag{4}$$

The encoder structure of the teacher generator G_T and the encoder structure of the student generator G_S are kept the same, The decoder structure of the teacher generator G_T and the decoder structure of the student network G_S correspond to each other, and they are up-sampled by a 3×3 transposed convolutional layer with a step size of two. The generated transformed image of the teacher generator G_T is denoted as p_t, and the generated transformed image of the student generator G_S is denoted as p_s. p_t is supervised by an online knowledge distillation loss on p_s. In order to enable the student generator G_S to acquire a priori knowledge of the G_T features, an online knowledge distillation loss is used to measure the difference between p_t and p_s. Structural similarity loss is a loss function based on the perceptual image quality loss function for evaluating and optimizing image quality in image generation and restoration tasks. The model training is guided by measuring the structural similarity between two images by:

$$\mathcal{L}_{SSIM}\left(p_t, p_s\right) = \frac{\left(2\mu_t\mu_s + C_1\right)\left(2\sigma_{ts} + C_2\right)}{\left(\mu_t^2\mu_s^2 + C_1\right)\left(\sigma_t^2 + \sigma_s^2 + C_2\right)} \tag{5}$$

where μ_t, μ_s are the estimated mean, σ_t^2, σ_s^2 are the standard deviation of contrast, and σ_{ts} is the covariance of the structural similarity estimate. C_1, C_2 are constants to avoid a zero denominator.

Feature loss $\mathcal{L}_{feature}$ evaluates image quality by calculating the distance between two image features in feature space. The goal is to capture high-level image features to ensure that the resulting images are similar not only at the pixel level, but also in terms of feature statistics. To ensure p_t and p_s have similar feature representations, we minimize the following feature loss:

$$\mathcal{L}_{feature}\left(p_t, p_s\right) = \frac{1}{C_j H_j W_j} \parallel \phi_j\left(p_t\right) - \phi_j\left(p_s\right) \parallel_1 \tag{6}$$

where $\phi_j\left(x\right)$ is the feature map obtained after processing the j layer of the network for the input x. The feature map is the dimension of the network. $C_j \times H_j \times W_j$ is the dimension of $\phi_j\left(x\right)$.

We use style loss $\mathcal{L}_{style}$ to penalize differences in style features such as color, texture, shape, etc.

$$\mathcal{L}_{style}\left(p_t, p_s\right) = \parallel G_j^\phi\left(p_t\right) - G_j^\phi\left(p_s\right) \parallel_1 \tag{7}$$

where $G_j^\phi\left(x\right)$ is the Gram matrix of layer j activations in the VGG network. This matrix captures the stylistic features of an image by computing the inner product relationship between channels in the activation map and is commonly used for tasks such as style migration. In this paper, the student generator G_S is optimized under the guidance of the teacher generator G_T. The G_S is trained and optimized without the involvement of the discriminator to obtain a better performing lightweight G_S network model.

Therefore, the total loss function and optimization objective are defined as:

$$\mathcal{L}_{KD}\left(p_t, p_s\right) = \lambda_{SSIM}\mathcal{L}_{SSIM} + \lambda_{feature}\mathcal{L}_{feature} + \lambda_{style}\mathcal{L}_{style} \qquad (8)$$

where λ_{SSIM}, λ_{feature}, and λ_{style} are hyperparameters used to balance the above loss terms.

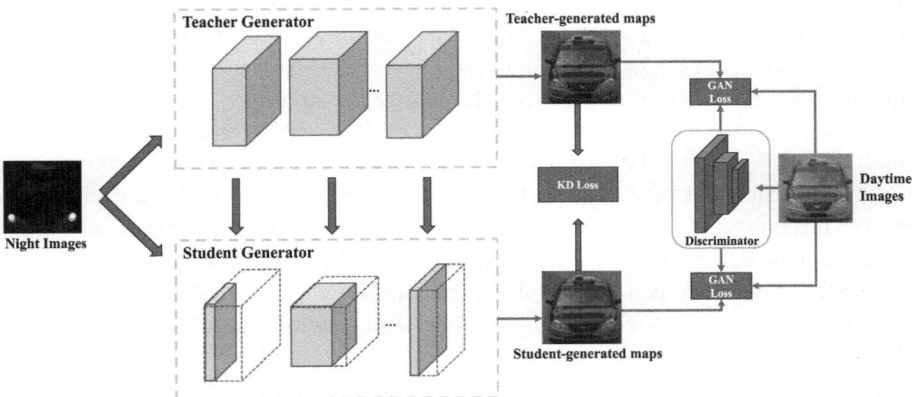

Fig. 3. Online Distillation Framework. The student generator G_S is optimized using only the teacher generator and G_T can be trained in the discriminator. Transferring knowledge from the teacher network to the student generator G_S ensures that key knowledge can be effectively transferred, thus helping the student generator to better understand and learn the task, training a lightweight student generator model, and using the student generator as a generator of images, which solves the problem of high complexity of the original generator

4   Experimentation

4.1   Experimental Settings

The Lightweight Attention-CycleGAN for Nighttime-Daytime Image Transformation is implemented on a PC with the following specifications: CPU model AMD Ryzen 7 7735H, GPU model GTX4060, and video memory size of 16 GB. The programming language used is Python, version 3 .7. The proposed deep learning network is based on CycleGAN, using the PyTorch framework. Training is performed using CycleGAN distillation method with a shared attention architecture for both teacher and student models, and the number of generator filters is 64 for the teacher model and 16 for the student model. The training process consists of 100 initial epochs and 100 learning rate decay epochs, and the number of discriminators is 4. The weights of the AGDs are 10, 10000, 10, and 0.00001 to control the importance of different loss terms.

In this paper, training is performed on the Dark Zurich training set, which contains 3044 daytime images and 2419 nighttime images randomly selected images within the dataset as test maps.

4.2 Metrics and Score Comparisons

The results of each module were objectively compared using six image assessment metrics, including four no-reference image quality metrics and two model size metrics. "BRISQUE" is defined as a metric that determines how close an image is to a natural scene by evaluating the amount of distortion in the image; the lower the score, the better the image quality is. "NIQE" examines the average features of a clean, undistorted image and compares these features with those of a test image. The closer the match between the features of the test image and the reference clean image means the higher the quality; "CEIQ" is a quality metric that addresses contrast distortion, the higher the score, the better the quality of the image; Michelson provides a quantitative measure of the degree of differentiation between bright and dark areas. Higher contrast values indicate greater intensity differences, leading to sharper boundaries and improved visual perception of details in the image; MAC is the total number of multiply-add cumulative operators in the model, where the lower the MAC value means the less complex the model. FID is the number of parameters in the network, where the lower the FID value means the less complex the model.

Table 1 compares the existing methods Pix2Pix, CycleGAN, GAN- Compression [26] and OMGD [27]. Observations show that the images generated by the model obtained better results in terms of Contrast Distortion Measure (CEIQ) and Picture Distortion (BRISQUE) compared to the existing methods. In addition, compared to existing methods, the proposed method shows superior scores in model size comparison, achieves model lightweighting, and somewhat balances the relationship between the quality of the generated images and the size of the model. Thus, the method in this paper improves the clarity of the objects in the image, improves the overall quality of the image, and demonstrates better lightweighting properties. This is important for improving the performance of image recognition systems for edge devices in nighttime environments.

Therefore, the proposed method enhances the local contrast and effectively solves the distortion problem of image in the process of night-to-day image transformation. Therefore, it improves the clarity of the objects in the image and the overall quality of the image. By the on-line knowledge distillation, it is compressed to obtain the student generator with less model complexity and shows better lightweight characteristics. That can be well applied to the edge devices in

Table 1. Comparison of average image assessment scores (up arrow: higher is better; down arrow: lower is better).

Method	BRISQUE↓	NIQE↓	CEIQ↓	Michelson contrast↑	MACs↓	Parameters↓
Pix2Pix	20.755	3.617	3.006	0.969	11.4M	56.80G
CycleGAN	19.879	3.814	3.974	4.061	11.30M	56.80G
GAN-Compression	21.635	3.950	4.125	3.651	0.70M	4.81G
OMGD	21.554	3.921	4.041	3.872	0.137M	1.408G
Ours	**19.562**	3.763	**3.962**	3.936	0.137M	1.408G

low illuminated scenes edge devices to realize night-to-day image transformation, which is important for improving the performance of detection and recognition models.

5   Conclusion

This paper has presented a shared attention network based on the attention mechanism, built on top of the CycleGAN structure. It also designs an online knowledge distillation method to compress and optimize the model, so as to obtain a lightweight model to achieve the cross-domain conversion of images. By extensive experiments and comparison with traditional methods, it is demonstrated that the proposed method can be used in a more lightweight model to ensure the quality of the generated images, after using it for night-time to day-time image conversion. Thus, it is of great significance to improve the performance of the image recognition system for edge devices in night-time environment. This is important for improving the performance of edge device image recognition systems in nighttime environments and provides a viable solution for deploying nighttime-daytime image transformation on edge devices.

References

1. Pang, Y., Lin, J., Qin, T., Chen, Z.: Image-to-image translation: methods and applications. IEEE Trans. Multimedia **24**, 3859–3881 (2021)
2. Goodfellow, I., et al.: Generative adversarial nets. Adv. Neural Inform. Process. Syst. **27** (2014)
3. Radford, A., Metz, L., Chintala, S.: Unsupervised representation learning with deep convolutional generative adversarial networks. arXiv preprint arXiv:1511.06434 (2015)
4. Isola, P., Zhu, J.-Y., Zhou, T., Efros, A.A.: Image-to-image translation with conditional adversarial networks. In: Proceedings of the IEEE Conference on Computer Vision and Pattern Recognition, pp. 1125–1134 (2017)
5. Howard, A.G., et al.: Mobilenets: efficient convolutional neural networks for mobile vision applications. arXiv preprint arXiv:1704.04861(2017)
6. Creswell, A., White, T., Dumoulin, V., Arulkumaran, K., Sengupta, B., Bharath, A.A.: Generative adversarial networks: an overview. IEEE Signal Process. Mag. **35**(1), 53–65 (2018)
7. Liu, G., Tang, H., Latapie, H.M., Corso, J.J., Yan, Y.: Cross-view exocentric to egocentric video synthesis. In: Proceedings of the 29th ACM International Conference on Multimedia, pp. 974–982 (2021)
8. Mirza, M., Osindero, S.: Conditional generative adversarial nets. arXiv preprint arXiv:1411.1784 (2014)
9. Perarnau, G., Van De Weijer, J., Raducanu, B., Álvarez, J.M.: Invertible conditional gans for image editing. arXiv preprint arXiv:1611.06355 (2016)
10. Tang, H., Xu, D., Liu, G., Wang, W., Sebe, N., Yan, Y.: Cycle in cycle generative adversarial networks for keypoint-guided image generation. In Proceedings of the 27th ACM International Conference on Multimedia, pp. 2052–2060 (2019)

11. Tang, H., Wang, W., Xu, D., Yan, Y., Sebe, N.: Gesturegan for hand gesture-to-gesture translation in the wild. In: Proceedings of the 26th ACM international conference on Multimedia, pp. 774–782 (2018)

12. Tang, H., Xu, D., Sebe, N., Wang, Y., Corso, J.J., Yan, Y.: Multi-channel attention selection gan with cascaded semantic guidance for cross-view image translation. In: Proceedings of the IEEE/CVF Conference on Computer Vision and Pattern Recognition, pp. 2417–2426 (2019)

13. Wang, T.-C., Liu, M.-Y., Zhu, J.-Y., Tao, A., Kautz, J., Catanzaro, B.: High-resolution image synthesis and semantic manipulation with conditional gans. In: Proceedings of the IEEE Conference on Computer Vision and Pattern Recognition, pp. 8798–8807 (2018)

14. Zhu, J.-Y., Park, T., Isola, P., Efros, A.A.: Unpaired image-to-image translation using cycle-consistent adversarial networks. In: Proceedings of the IEEE International Conference on Computer Vision, pp. 2223–2232 (2017)

15. Kim, T., Cha, M., Kim, H., Lee, J.K., Kim, J.: Learning to discover cross-domain relations with generative adversarial networks. In: International Conference on Machine Learning, pp. 1857–1865. PMLR (2017)

16. Xu, D., Wang, W., Tang, H., Liu, H., Sebe, N., Ricci, E.: Structured attention guided convolutional neural fields for monocular depth estimation. In: Proceedings of the IEEE conference on computer vision and pattern recognition, pp. 3917–3925 (2018)

17. Liang, X., Zhang, H., Xing, E.P.: Generative semantic manipulation with contrasting gan. arXiv preprint arXiv:1708.00315 (2017)

18. Bahdanau, D.: Neural machine translation by jointly learning to align and translate. arXiv preprint arXiv:1409.0473 (2014)

19. Vaswani, A.: Attention is all you need. Adv. Neural Inform. Process, Syst (2017)

20. Alexey, D.: An image is worth 16x16 words: Transformers for image recognition at scale. arXiv preprint arXiv: 2010.11929 (2020)

21. Shen, Z., Zhang, M., Zhao, H., Yi, S., Li, H.: Efficient attention: attention with linear complexities. In: Proceedings of the IEEE/CVF Winter Conference on Applications of Computer Vision, pp. 3531–3539 (2021)

22. Romero, A., Ballas, N., Kahou, S.E., Chassang, A., Gatta, C., Bengio, Y.: Fitnets: Hints for thin deep nets. arXiv preprint arXiv:1412.6550 (2014)

23. Jung, C., Kwon, G., Ye, J.C.: Exploring patch-wise semantic relation for contrastive learning in image-to-image translation tasks. In: Proceedings of the IEEE/CVF Conference on Computer Vision and Pattern Recognition, pp. 18260–18269 (2022)

24. Chen, P., Liu, S., Zhao, H., Jia, J.: Distilling knowledge via knowledge review. In: Proceedings of the IEEE/CVF Conference on Computer Vision and Pattern Recognition, pp. 5008–5017 (2021)

25. Park, W., Kim, D., Lu, Y., Cho, M.: Relational knowledge distillation. In: Proceedings of the IEEE/CVF Conference on Computer Vision and Pattern Recognition, pp. 3967–3976 (2019)

26. Li, M., Lin, J., Ding, Y., Liu, Z., Zhu, J.-Y., Han, S.: Gan compression: efficient architectures for interactive conditional gans. In: Proceedings of the IEEE/CVF Conference on Computer Vision and Pattern Recognition, pp. 5284–5294 (2020)

27. Ren, Y., Wu, J., Xiao, X., Yang, J.: Online multi-granularity distillation for gan compression. In: Proceedings of the IEEE/CVF International Conference on Computer Vision, pp. 6793–6803 (2021)

Generative Image Steganography Based on Latent Space Vector Coding and Diffusion Model

Weisong Liu[1] , Jianfeng Yang[2], Jiewen Huang[2], Ahmed A. Abd El-Latif[3], and Zhili Zhou[1(✉)]

[1] School of Artificial Intelligence, Guangzhou University, Guangzhou, China
zhou_zhili@163.com
[2] School of Computer Science and Cyber Engineering, Guangzhou University, Guangzhou, China
[3] EIAS Data Science Lab, College of Computer and Information Sciences, and Center of Excellence in Quantum and Intelligent Computing, Prince Sultan University, Riyadh 11586, Saudi Arabia

Abstract. Image steganography is a technology that embed secret information within a cover image to obtain a stego-image for covert communication. The transmission of undetectable stego-images via social media can facilitate secure and efficient covert communication. Recently, the generative image steganography has achieved promising performance with the rapid development of generative models. However, existing generative image steganography suffers from issues, i.e., the low quality of the generated stego-image and the limited hiding capacity. To address these issues, this paper proposes a generative image steganography scheme based on latent space vector coding and diffusion model. It consists of an optimized diffusion denoising network, the latent space vector coding mapping network, and the information extraction network. In this scheme, the secret information is encoded as a latent space vector, which is then transformed in the input diffusion model to generate the stego-image. Experimental results demonstrate that this scheme achieves superior performance compared with existing generative image steganography methods.

Keywords: Generative image steganography · Diffusion model · Space vector

1 Introduction

Image steganography [1,2] is the technology that conceals secret information within an image without arousing notice. In the field of traditional image steganography, researchers utilize the inherent redundancy of cover image, making only slight modifications to them, with the objective of embedding secret information into the cover images in a manner that is undetectable. However,

F. Zhang et al. (Eds.): AIS&P 2024, LNCS 15399, pp. 167–179, 2025.
https://doi.org/10.1007/978-981-96-1148-5_14

modifying the image will inevitably leave the modification traces into the cover image, which may be identified and used by steganalysis tools to successfully detect the existence of the secret message. To resist steganalysis detection, recently, researchers have proposed the concept of generative steganography [3–5]. In contrast to the traditional steganography, generative steganography directly generates a stego-image driven by the secret information for covert communication. As no modification is left during the steganography process, generative steganography can effectively resist the steganalysis detection. Consequently, generative steganography has attracted a lot of attention from in the field of information hiding.

However, for generative image steganography, although it is inherently not straightforward for machines to detect whether it contains secret information or not, it would arouse the suspicion by human eyes, and thus the quality of the generated stego-image is also very important. High-quality stego-image will greatly improve the security of image steganography. This paper proposes a generative image steganography method based on latent space vector coding and diffusion model. The aim of this method is to enhance the quality of stego-images while increasing the steganographic capacity to satisfy the requirement of practical applications in real-world scenarios. It consists of an optimized diffusion denoising network, a latent space vector coding mapping network, and an information extraction network. The latent space vectors can be input in the diffusion model to generate a high-quality stego-image with large hiding capacity. The experimental results demonstrate that the proposed method shows superior performance compared to existing generative image steganography methods.

2 Preliminaries and Related Work

2.1 Modified Image Steganography

The modification-based image steganography methods involve the selection of an appropriate existing cover image, followed by the embedding of secret information by slight modifications to the cover image. Filler and Fridrich et al. [6] proposed a steganography framework based on STC (Syndrome-Trellis Codes). This framework defines a modification cost function based on the statistical properties of the image, and minimizes the overall distortion cost via steganographic coding techniques, so that the distortion of the cover image can be controlled as small as possible. Tang et al. [7] proposed a GAN-based automatic steganographic distortion learning scheme, ASDL-GAN. Yang et al. [8] further optimized the structure of GAN for steganography. Furthermore, the UT-GAN(An Embedding Cost Learning Framework Using GAN) method has been proposed based on ASDL-GAN. This method uses the U-Net(Convolutional Networks for Biomedical Image Segmentation) [9] as a generator and designs a double-tanh activation function as a simulated embedder. This results in an improvement in the performance of modifying the cost function. Guan et al. [10] also proposed the DeepMIH steganography architecture. This architecture realizes the hiding of multiple images in one loaded dense image under low distortion constraints. The proposed architecture allows for the hiding of a single image at a time, utilizing the potential remaining image hiding space through multiple loops, to minimize image distortion and achieve high-capacity embedding while maintaining the visual quality of the stego-image. The modification-based image steganography has the ability of embedding the secret information with less image distortion. However, the presence of modification traces poses a significant challenge to resist the steganalysis detection.

2.2 Generative Image Steganography

The advantage of generative steganography is that it does not require any modification of the cover image. This makes the stego-image difficult to be detected by steganalysis tools. Zhou et al. [3] proposed a framework for generative steganography based on image feature matching. This framework selects a series of natural images from an image database that already contain or can represent secret information. These natural images are then treated as the stego-images. However, the steganographic capacity is very low and impractical for use in real-world steganography tasks. Wu et al. [4] employed steganographic methods during texture image synthesis. The method selects the corresponding texture image blocks according to the secret information and stitches them into a whole texture image, which serves as the stego-image. Liu et al. [11] proposed a steganography method based on Image Disentanglement Autoencoder for Steganography (IDEAS). The method transforms the secret information into the structural characteristics of the image, which are then utilized as input for the autoencoder to generate the stego-image. However, these steganography methods based on GANs or self-encoders are unable to accurately extract the secret information from the generated images due to the inherent non-reversibility of the network structure of GANs or self-encoders. Zhou et al. [5] employed the GLOW model [12] to develop the Secret-to-Image Reversible Transformation (S2RT) method, which utilizes the secret information to transform the image. Sequentially, a steganography method based on the GLOW model [12] and Secret-to-Image Reversible Transformation (S2IRT) has been proposed [5], which has the potential to significantly enhance steganographic capacity while ensuring the accurate extraction of secret information. The S2IRT method has the potential to significantly enhance steganographic capacity while ensuring the extraction accuracy of the secret information. However, S2IRT is constrained in terms of visual quality and the diversity of steganographic images.

2.3 Diffusion Model

The Diffusion Model (DM) [13] is a generative model designed for the generation of new data samples, including those of textual, visual, or auditory nature. The model is based on the theoretical framework of the diffusion process, which implements the generation of novel data samples through the incremental introduction of random noise to the original data set, followed by the learning of the distribution of the original data set from the noise. For image generation, the diffusion model's generation principle can be regarded as a denoising operation applied to an input image. The model can be trained to learn how to denoise this input image in a meaningful manner. Consequently, the model can further comprehend how to generate meaningful and high-quality images. The computational resources required for the diffusion model to generate new data samples are greater than those required by other generative models. This is due to the necessity of multiple iterations in each denoising process, which results in a longer generation time. Moreover, the model necessitates a substantial quantity of training data to facilitate the learning of the data distribution. Nevertheless, due to its excellent generative ability and stable performance, the diffusion model continues to receive extensive attention in the field of artificial intelligence.

3   Proposed Approach

This section presents an overview of proposed generative image steganography framework, which is based on the latent space vector coding and diffusion model. The framework consists of three principal network structures: the optimized diffusion denoising network, the latent space vector coding mapping network, and the latent space vector extraction network. In this method, the optimized diffusion denoising network is initially devised by optimizing the conventional diffusion model to meet the requirements of the generative image steganography methodology based on latent space vectors. Subsequently, an INN-based cryptospace vector coding and decoding mapping network is designed. This enables the network model and secret information to be mapped into latent vectors, which are then utilized for the generation of the subsequent stego-image. Furthermore, the network is reversible, thereby facilitating the accurate extraction of the secret information during the reversal process. In conclusion, the method incorporates a latent space vector extraction network, based on an encoder and decoder structure, which primarily transforms the stego-images into latent space vectors for subsequent operations. The general framework diagram of the method is illustrated in Fig. 1.

3.1   Optimize Diffusion Denoising Network

This subsection is dedicated to an in-depth examination of the optimized diffusion denoising network, which serves as the fundamental generative model in this framework. In the original diffusion network model, the forward diffusion process for the latent space vector x_T is a Markov process, whereby noise should be added continuously. The forward diffusion process begins with an image and proceeds through a series of time steps, during which noise is continuously added, ultimately resulting in a completely noisy image. This can be represented by

$$q\left(x_{1:T} \mid x_0\right) = \prod_{t=1}^{T} q\left(x_t \mid x_{t-1}\right), q\left(x_t \mid x_{t-1}\right) = \mathcal{N}\left(x_t; \sqrt{1 - \beta_t}x_{t-1}, \beta_t I\right) \quad (1)$$

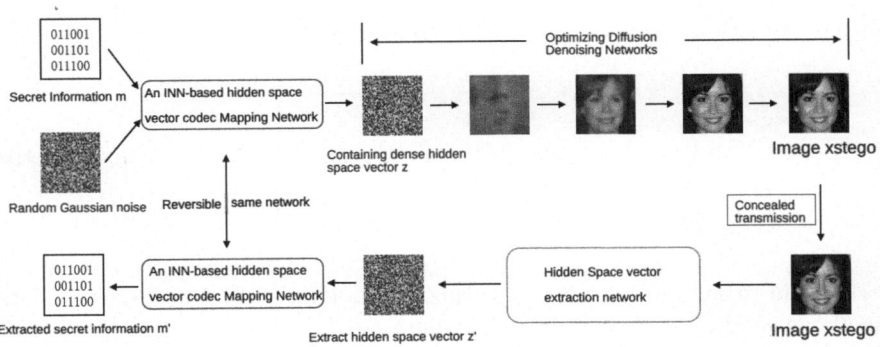

Fig. 1. A general framework for generative image steganography based on latent space vector coding and diffusion model

In the context of the diffusion model, the parameter T represents the number of steps that the model passes through in a single process. It can be observed that, in general, an increase in the number of steps results in a more pronounced generation effect, although this is accompanied by a longer processing time. Conversely, a reduction in the number of steps leads to a more general generation effect. The parameter β_t is a preset parameter. $\beta_t = \left(1 - \frac{\alpha_t}{\alpha_{t-1}}\right)$.

In the case of the sample image x and the latent space vector z, the complete forward noise addition and reverse denoising procedure can be represented by

$$x = x_0 \leftrightarrow x_1 \leftrightarrow x_2 \leftrightarrow \cdots \leftrightarrow x_{T-1} \leftrightarrow x_T = z \tag{2}$$

In the context of the entire diffusion network model, the term "forward" refers to the process of converting the sample image, denoted by x, into a latent space vector z, in a gradual and systematic manner. The inverse process is the transformation of the latent space vector z into the sample image x. This process can be considered equivalent to the generation of an image.

In regard to the forward process, each of these minor stages can be represented in an alternative format as follows:

$$
\begin{aligned}
x_t &= \alpha_t x_{t-1} + \beta_t \varepsilon_t \\
&= \alpha_t \left(\alpha_{t-1} x_{t-2} + \beta_{t-1} \varepsilon_{t-1}\right) + \beta_t \varepsilon_t \\
&= \alpha_t \left(\alpha_{t-1} \left(\alpha_{t-2} x_{t-3} + \beta_{t-2} \varepsilon_{t-3}\right) + \beta_{t-1} \varepsilon_{t-1}\right) + \beta_t \varepsilon_t \\
&= \cdots\cdots\cdots \\
&= \left(\alpha_t * \alpha_{t-1} * \cdots * \alpha_1\right) x_0 + \left(\alpha_t * \cdots * \alpha_2\right) \beta_1 \varepsilon_1 + \left(\alpha_t * \cdots * \alpha_3\right) \beta_2 \varepsilon_2 \\
&\quad + \cdots + \alpha_t \beta_{t-1} \varepsilon_{t-1} + \beta_t \varepsilon_t
\end{aligned}
\tag{3}
$$

The sum of terms with ε_t, such as $\left(\alpha_t * \cdots * \alpha_2\right) \beta_1 \varepsilon_1$, represents the sum of a number of mutually independent noises that obey a normal distribution. For brevity, we will denote $\left(\left(\alpha_t * \alpha_{t-1} * \cdots * \alpha_1\right)\right)$ as $\overline{\alpha_t}$. Then $\bar{\beta}_t = \sqrt{1 - \overline{\alpha_t}^2}$, thereby enabling the simultaneous derivation of both $\bar{\beta}_t$ and $\overline{\alpha_t}$ through the mutual deduction of the computations. In addition, Bayes' theorem states that:

$$q\left(x_{t-1} \mid x_1, x_0\right) = \frac{q\left(x_t \mid x_{t-1}\right) q\left(x_{t-1} \mid x_0\right)}{q\left(x_t \mid x_0\right)} \tag{4}$$

According to Bayes' theorem in the above equation, it can be deduced that:

$$q\left(x_{t-1} \mid x_t, x_0\right) = \mathcal{N}\left(x_{t-1}; \frac{\alpha_t \overline{\beta_{t-1}}^2}{\overline{\beta}_t^2} x_t + \frac{\overline{\alpha_{t-1}}^2 \beta_t^2}{\overline{\beta}_t^2} x_0, \frac{\overline{\beta}_{t-1}^2 \beta_t^2}{\overline{\beta}_t^2} I\right) \tag{5}$$

In order to compute x_{t-1} by x_t without relying on x_0, it is necessary to derive the above conditional probability distribution. Therefore, we make a prediction of x_0 by using a neural network model ϵ_ϑ , where the neural network model is U-Net. This process can be expressed as approximating the estimation of x_0 by using $\bar{\mu}\left(x_t\right)$. The expression of $\bar{\mu}\left(x_t\right)$ is as follows:

$$\bar{\mu}\left(x_t\right) = \frac{1}{\overline{\alpha}_t}\left(x_t - \sqrt{1 - \bar{\alpha}_t^2}\epsilon_\vartheta\left(x_t, t\right)\right) \tag{6}$$

At this point, the distribution $q\left(x_{t-1} \mid x_t, x_0 \approx \bar{\mu}\left(x_t\right)\right)$ of the generated image can be calculated with the following expression:

$$q\left(x_{t-1} \mid x_t, x_0 \approx \bar{\mu}(x_t)\right) = \sqrt{\overline{\alpha_{t-1} * \widehat{x_0}_t}} + \sqrt{1 - \overline{\alpha_{t-1}} - \sigma_t^2}\varepsilon_\theta + \sigma_t\varepsilon_{random} \tag{7}$$

In this context, the variable $\widehat{x_0}_t$ denotes the current step image that has been directly predicted using the input variable x_t. The term ε_ϑ represents the noise that has been predicted by the neural network model, while the variable ε_{random} denotes the noise that has been introduced during the process as a result of random sampling. The variable σ_t is a controllable parameter.

Subsequently, for the generative image steganography method based on latent space vectors, this method sets σ_t to 0, thereby yielding the following expression for the distribution of the generated image:

$$q\left(x_{t-1} \mid x_t, x_0 \approx \bar{\mu}\left(x_t\right)\right) = \sqrt{\overline{\alpha_{t-1} * \widehat{x_0}}} + \sqrt{1 - \overline{\alpha_{t-1}}}\varepsilon_\vartheta \tag{8}$$

At this juncture, the optimized diffusion denoising network will not introduce uncontrollable random noise during the generation of the image. Instead, it will only add noise calculated by the weight parameters of the neural network model, which is ε_ϑ. This noise can be controlled. Accordingly, the latent space vector z can be manipulated to generate the corresponding densely loaded image x_{stego} .

In order to train the optimized diffusion denoising network, which is derived from the above equation, it is necessary to predict the image x_0 by the neural network ϵ_ϑ. Consequently, the loss function $Loss$ is defined as follows:

$$\text{Loss} = \left\|x_0 - \bar{\mu}\left(x_t\right)\right\|^2 = \frac{\overline{\beta}_t^2}{\overline{\alpha}_t^2}\left\|\varepsilon - \epsilon_\vartheta\left(\overline{\alpha}_t x_0 + \overline{\beta}_t\varepsilon, t\right)\right\|^2 \tag{9}$$

The optimized diffusion denoising network employs a loss function, $Loss$, to minimize the loss during the training process. This enables the continuous optimization of weighting parameters, thus facilitating the identification of optimal network parameters and the generation of a high-quality generative steganographic cryptographically-loaded image, x_{stego} .

3.2 Latent Space Vector Encoding and Decoding Mapping Network Based on INN

This subsection focuses on the INN-based latent space vector codec mapping network, which is mainly based on the reversible neural network INN [14]. The network is reversible and is capable of converting two inputs, namely the latent space vector z and the secret message m, which are randomly sampled from a Gaussian distribution, into the latent space vector, i.e., Z_{stego}. Concurrently, the input model of Z_{stego} is also reversibly converted into the latent space vector z and the secret message m', thereby facilitating the extraction of the secret message.

The following section will briefly introduce the reversible neural network, INN [14], which is comprised of a N-structure of consecutive reversible blocks that are capable of performing both forward and reverse computation operations. The formula for its forward computation is given as follows:

$$y_1 = x_1 + f(x_2)$$
$$y_2 = x_2 + g(y_1) \tag{10}$$

The formula for calculating the inverse is given as follows:

$$x_2 = y_2 - g(y_1)$$
$$x_1 = y_1 - f(y_2) \tag{11}$$

The process diagrams for the forward and reverse computation of the reversible neural network INN are illustrated in Fig. 2.

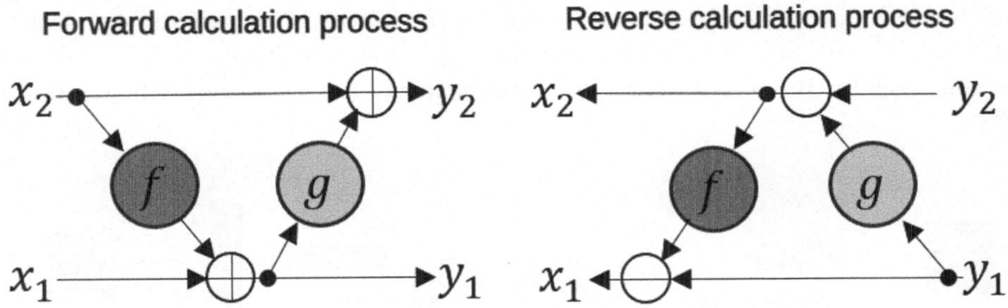

Fig. 2. Reversible block computation procedure for reversible neural network INNs

The combination of the two formulas for forward and inverse computation, along with the Jacobi determinant computation, allows each reversible block to perform reversible computation operations. Furthermore, as the reversible neural network (INN) consists of N reversible blocks, the network is also capable of executing reversible computational operations.

The principal configuration of the INN-based latent space vector codec mapping network, as employed in this methodology, is illustrated in Fig. 3.

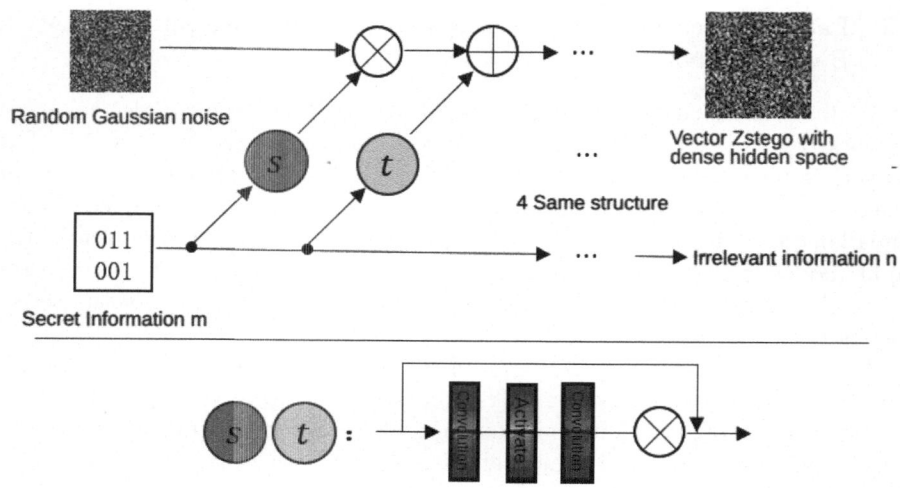

Fig. 3. The main structure of INN-based latent space vector codec mapping network

Figure 3 describes solely the forward process of the network; however, the network can be also converted into a reverse operation, whereby the forward direction of the opposite calculation can be performed. The reversible network allows for the simultaneous generation of the latent space vector, i.e., z_{stego} , and extraction of the secret information m'.

3.3 Latent Space Vector Extraction Network

This subsection will present the design of the latent space vector extraction network. The method is modeled after the U-Net network structure, which serves as the basis for its design. The overall structure of the network is illustrated in Fig. 4.

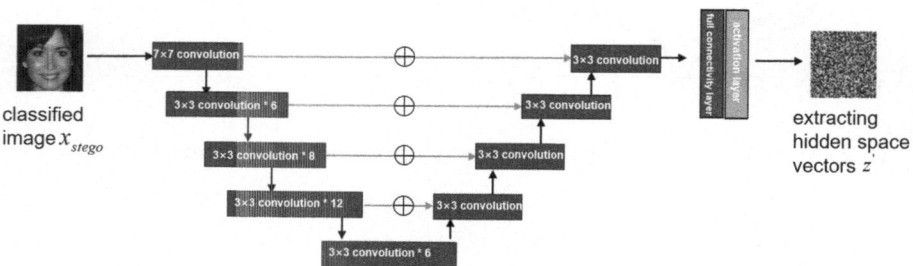

Fig. 4. The structure of latent space vector extraction network

As illustrated in Fig. 4, the latent space vector extraction network is a symmetric network structure, and its overall architecture is based on the U-Net network. The network comprises an encoder on the left and a decoder on the right. The encoder is responsible for down-sampling operations, while the decoder is tasked with up-sampling operations. The encoder and decoder are composed of multiple convolutional layers. The convolutional layers of the encoder are primarily based on those of the residual network, i.e., ResNet-34. Given the remarkable performance of ResNet-34, this paper draws upon the highly effective modules of this network in the design of the latent space vector extraction network. Following the input of the image into the network, the dimension will undergo continuous reduction during the downsampling process of the network encoder, while the dimension can be enlarged during the upsampling process. Concurrently, residual structures are linked to the convolutional layers in the same dimension, thereby enhancing the network's performance. The input stego-image, i.e., x_{stego} , can be transformed into a secret-containing latent space vector z' of the same dimension by the latent space vector extraction network for extracting the secret information.

4 Experimental Results

This section will present the experimental details of the proposed generative image steganography method. It includes the description of image dataset used for training the model in the experiments, the parameter settings of the model and the algorithm, as well as the evaluation of the performance of proposed method. The results of the experiments will also be analyzed in detail.

4.1 Dataset and Experimental Environment Settings

The effectiveness of the proposed method is evaluated on four high-definition real-image datasets, namely the CelebA, FFHQ, Animals-10, and LSUN datasets. The diversity of these image datasets encompasses faces, animals, and buildings, which enables comprehensive and effective evaluation of the generation performance of the proposed method. Furthermore, these datasets are also commonly employed in existing image generation methods.

Furthermore, the datasets are preprocessed by aligning and cropping images to a uniform size of 256*256. Additionally, a small number of images of the same kind with significant differences were eliminated by manual screening.

All experiments in this section were conducted on the NVIDIA RTX 3090 GPU platform and programmed using the PyTorch framework and the Python language.

4.2 The Security Evaluation of Generative Image Steganography

To evaluate the steganographic security of the generative image steganography method based on the latent space vector coding and diffusion model, this subsection presents an evaluation of the security in two separate aspects: 1) the security evaluation of resisting steganography analysis, and 2) the security evaluation of the quality of generated stego-image.

(1) Security Evaluation Against Steganography Analyzers This subsection assesses the capacity of the method to withstand steganalysis, a metric that directly correlates with the security of the steganographic method. The quantitative metric for estimating the ability of the method to resist a steganalysis attack is also employed as a performance estimate (P_E), which is calculated using the following formula:

$$P_E = \min_{P_{FA}} \frac{1}{2} (P_{FA} + P_{MD}) \tag{12}$$

The probability of false detection for a steganalyzer is represented by P_{FA}, while the probability of missed detection is denoted by P_{MD}. As the formula indicates, an elevated P_E value signifies enhanced resilience to detection by steganalyzers.

In this subsection, two steganalyzers, namely SRM and XuNet, are employed to ascertain whether an image in question contains any secret information. In this study, 5,000 images devoid of any embedded secret information and an additional 5,000 randomly generated images lacking any hidden secret information were manually selected. Subsequently, 10,000 stego-images containing hidden secret information were generated by various generative image steganography techniques. The total number of these 20,000 images were employed for the training of the steganalyzer. Following the training of the steganalyzer, this paper presents a comparison of the experimental results obtained with this method and those of other generative image steganography methods. Table 1 illustrates the steganalysis resistance of the various methods in relation to SRM and XuNet.

Table 1. P_E of different methods for SRM and XuNet

cryptographic analyzer	norm	Deep-Stego	SWE	IDEAS	S2IRT	This method
SRM	P_E	0.328	0.296	0.283	0.290	0.287
XuNet	P_E	0.319	0.301	0.313	0.251	0.320

Table 1 demonstrates that this method achieves a slight advantage over similar techniques in terms of resisting steganalysis. This is particularly evident when targeting XuNet, where this method demonstrates superior performance compared to other techniques. However, in general, these comparisons of generative image steganography do not yield significant differences in the performance of resisting steganalysis. Given that generative image steganography is inherently difficult to detect by steganalysis, it is essential to prioritize the visual quality and diversity of the stego-images it generates.

(2) Security Evaluation for Generated Stego-Image Quality. The security of image steganography is dependent upon the utilisation of generative image steganography, which is capable of resisting detection by typical steganalyzers.

In addition to this, the visual imperceptibility of the steganographic image is of significant importance for the security of image steganography. The imperceptibility of a steganographic image is contingent upon the quality of the steganographic image itself. In order to assess the quality of the generated stego-images produced by a variety of generative image steganography methods with differing steganographic capacities, we employed the use of FID, a quantitative measure. The results of this assessment are presented in Table 2, which lists the FID values of different generative image steganography methods under specified steganographic capacity and across a range of image datasets.

Table 2. FID of different generative image steganography methods under different image datasets

Method	FFHQ	CelebA	Animals-10	LSUN	FID mean
IDEAS	29.02	27.13	25.91	17.20	25.1±9.7
SWE	149.37	136.22	109.11	96.21	100.39±50.1
S2IRT	62.59	54.22	48.60	49.99	49.61±15.9
This method	17.81	19.26	22.87	25.14	25.72±8.2

As evidenced by the experimental results presented in Table 2, this method demonstrates superior performance on the majority of datasets, including FFHQ, CelebA, and Animals-10. It achieves the lowest FID score, with only a slight discrepancy compared to the IDEAS method on the LSUN dataset within the architectural scene category. This suggests that this method excels in generating high-quality stego-images. While the IDEAS method does not achieve the lowest FID score on the LSUN dataset, it has a relatively limited steganographic capacity. In contrast, the present method has a significantly larger steganographic capacity, which makes it a more versatile option for various applications. Furthermore, it is postulated that this phenomenon may also be attributable to an inherent error in the training model. In order to optimise the training of the diffusion model, it is necessary to increase the size of the training dataset and the number of images included. In order to train the diffusion model, a larger dataset and more images are required as the training set. In contrast, the GAN-based method needs a relatively smaller number of images as the training set, yet still produces satisfactory generation results.

4.3   Evaluation of Information Extraction Rate

The experiments presented in this subsection are designed to assess the efficacy of the proposed method in extracting secret information, by comparing its performance with that of alternative image steganography techniques at varying levels of steganographic capacity. The results of this comparison are illustrated in Table 3.

Table 3. Information extraction rate of different image steganography methods at different steganographic capacities

Method	0.1	0.2	0.5	2.0	4.0
Deep-Stego	98.52%	96.21%	95.92%	overcapacity	overcapacity
SWE	78.51%	71.89%	70.07%	69.02%	overcapacity
IDEAS	96.23%	overcapacity	overcapacity	overcapacity	overcapacity
S2IRT	100%	100%	100%	100%	99.48%
This method	100%	100%	100%	99.25%	98.71%

The data presented in the Table 3 demonstrates that when the hidden write capacity falls within the range of $0.1 \leq bpp \leq 4.0$, all the methods proposed in this chapter exhibit a high extraction rate and satisfactory performance. This is primarily attributable to the reversible property of the reversible neural network employed in the INN-based latent space vector codec mapping network, which enables the reversible extraction of secret information. However, when the hiding capacity is substantial, the extraction rate does not attain 100% due to the presence of a certain degree of extraction error in the latent space vector extraction network. Furthermore, the S2IRT method exhibits a higher extraction rate within the range of 2.0 to 4.0. This is primarily attributable to the fact that the method is built upon the diffusion model, which has superior reversible characteristics, thereby resulting in a higher extraction rate of secret information in the S2IRT method.

4.4   Limitation

We propose a generative image steganography method based on latent space vector coding as a solution to the existing problems of low-quality generated stego-images and small steganographic capacity. Nevertheless, the proposed method still exhibits shortcomings in its utilization of the diffusion model within the domain of generative image steganography, and there remains scope for further optimization. The diffusion model in the proposed method in this paper has yet to achieve reversible generation of stego-images and extraction of secret information. As the diffusion model is reversible, the performance of existing generative image steganography methods would be significantly enhanced. It is therefore recommended that further in-depth research will be carried out on the breakthrough point of the diffusion model.

5   Conclusion

This paper has presented a novel generative image steganography method based on the latent space vector coding and diffusion model. This method introduces a generative image steganography method based on the diffusion model, and the experimental results demonstrate that this method exhibits superior performance compared to existing generative image steganography methods.

References

1. Pevný, T., Filler, T., Bas, P.: Using high-dimensional image models to perform highly undetectable steganography. In: Böhme, R., Fong, P.W.L., Safavi-Naini, R. (eds.) IH 2010. LNCS, vol. 6387, pp. 161–177. Springer, Heidelberg (2010). https://doi.org/10.1007/978-3-642-16435-4_13
2. Holub, V., Fridrich, J.: Designing steganographic distortion using directional filters. In: 2012 IEEE International Workshop on Information Forensics and Security (WIFS), pp. 234-239. IEEE (2012)
3. Zhou, Z., Sun, H., Harit, R., Chen, X., Sun, X.: Coverless image steganography without embedding. In: Huang, Z., Sun, X., Luo, J., Wang, J. (eds.) ICCCS 2015. LNCS, vol. 9483, pp. 123–132. Springer, Cham (2015). https://doi.org/10.1007/978-3-319-27051-7_11
4. Wu, K.C., Wang, C.M.: Steganography using reversible texture synthesis. IEEE Trans. Image Process. **24**(1), 130–139 (2014)
5. Zhou, Z., Su, Y., Li, J., et al.: Secret-to-image reversible transformation for generative steganography. IEEE Trans. Dependable Secure Comput. **20**(5), 4118–4134 (2022)
6. Filler, T., Judas, J., Fridrich, J.: Minimizing additive distortion in steganography using syndrome-trellis codes. IEEE Trans. Inf. Forensics Secur. **6**(3), 920–935 (2011)
7. Tang, W., Tan, S., Li, B., et al.: Automatic steganographic distortion learning using a generative adversarial network. IEEE Signal Process. Lett. **24**(10), 1547–1551 (2017)
8. Yang, J., Ruan, D., Huang, J., et al.: An embedding cost learning framework using GAN. IEEE Trans. Inf. Forensics Secur. **15**, 839–851 (2019)
9. Ronneberger, O., Fischer, P., Brox, T.: U-Net: convolutional networks for biomedical image segmentation. In: Navab, N., Hornegger, J., Wells, W.M., Frangi, A.F. (eds.) MICCAI 2015. LNCS, vol. 9351, pp. 234–241. Springer, Cham (2015). https://doi.org/10.1007/978-3-319-24574-4_28
10. Guan, Z., Jing, J., Deng, X., et al.: DeepMIH: deep invertible network for multiple image hiding. IEEE Trans. Pattern Anal. Mach. Intell. **45**(1), 372–390 (2022)
11. Liu, X., Ma, Z., Ma, J., et al.: Image disentanglement autoencoder for steganography without embedding. In: Proceedings of the IEEE/CVF Conference on Computer Vision and Pattern Recognition, pp. 2303-2312 (2022)
12. Kingma, D.P., Dhariwal, P.: Glow: generative flow with invertible 1x1 convolutions. Adv. Neural Inform. Process. Syst. **31** (2018)
13. Ho, J., Jain, A., Abbeel, P.: Denoising diffusion probabilistic models. Adv. Neural. Inf. Process. Syst. **33**, 6840–6851 (2020)
14. Karami, M., Schuurmans, D., Sohl-Dickstein, J., et al.: Invertible convolutional flow. Adv. Neural Inform. Process. Syst. **32** (2019)